Leckie×Leckie

Scotland's leading educational publishers

ESSENTIAL EXAM SKILLS

Higher
PHYSICS
GRADE BOOSTER

Higher **PHYSICS** *GRADE BOOSTER*

John Irvine • Michael Murray

001/21082017

10 9 8 7 6 5 4 3 2 1

ISBN 9780007590858

Published by
Leckie & Leckie Ltd
An imprint of HarperCollinsPublishers
Westerhill Road, Bishopbriggs, Glasgow, G64 2QT
T: 0844 576 8126   F: 0844 576 8131
leckieandleckie@harpercollins.co.uk
www.leckieandleckie.co.uk

Commissioning editor: Clare Souza
Managing editor: Craig Balfour

Special thanks to
Louise Robb (copy edit and proofread)
Philip Bradfield (proofread)
Keren McGill (proofread)
Jouve India (layout and illustration)
Paul Oates (cover)

Printed in China

A CIP Catalogue record for this book is available from the British Library.

Acknowledgements
All images © Shutterstock.com

SQA questions reproduced with permission (solutions do not emanate from SQA), Copyright © Scottish Qualifications Authority.

Contents

Introduction

Who is this book for?

This book has been written to support candidates in obtaining a pass in CfE Higher Physics.

The purpose of this book

This book has been designed to help candidates in dealing with the common areas of difficulty within the CfE Higher Physics course. It also allows students to develop their ability in giving better-quality answers in the final examination.

What is in this book?

This book gives a description of the components of the Higher Physics course and the various assessments required for a course award.

This book lists the common areas of difficulty that prevent many students from scoring highly in the final examination. The book contains advice and examples to assist candidates in avoiding these common mistakes.

There are also sample questions with example answers to improve the ability of students to identify strong or weak answers.

Additionally there is a section providing advice on the completion of the course assignment.

How to use this book

This book should be used alongside your work in class and any summary notes or textbooks you might already have.

This book should be used throughout the year, as well as in the run up to prelim and final examinations. When preparing for exams, this book should be used in conjunction with past-paper questions to enable you to transfer any knowledge or skills gained into producing high-quality answers.

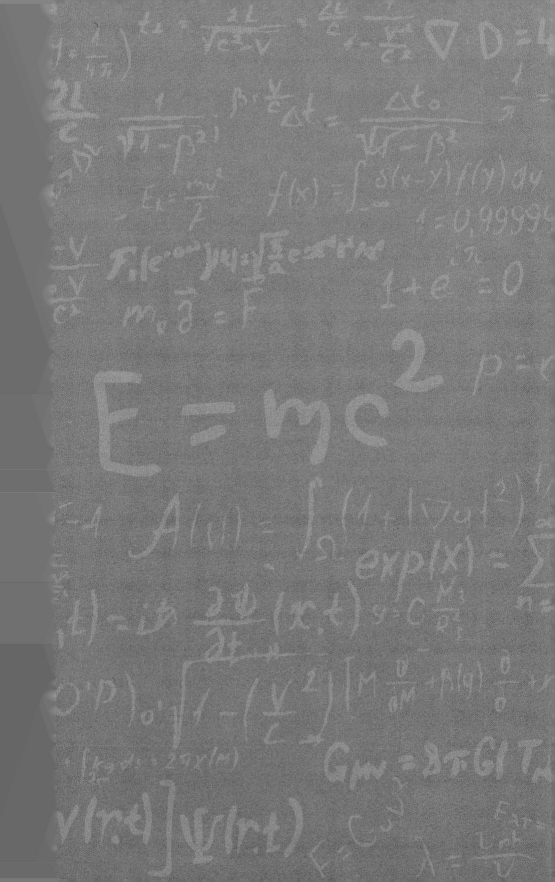

Higher Physics

Course content

Full details concerning the course content and the assessment specifications for both the final exam question paper and the assignment can be found on the Scottish Qualifications Authority Website: www.sqa.org.uk. The materials include course and unit support notes, specimen question papers with their mark schemes, general coursework information giving details of how your assignment will be marked, and the Physics relationships sheet. Although much of the information is written for teachers, you are allowed to download the information for your own use as a student.

The main aims of this course are for learners to:

- develop and apply knowledge and understanding of physics
- develop an understanding of the role of physics in scientific issues and relevant applications of physics, including the impact these could make on society and the environment
- develop scientific inquiry and investigative skills
- develop scientific analytical thinking skills, including scientific evaluation, in a physics context
- develop the use of technology, equipment and materials, safely, in practical scientific activities
- develop planning skills
- develop problem-solving skills in a physics context
- use and understand scientific literacy to communicate ideas and issues, and to make scientifically informed choices
- develop the knowledge and skills for more advanced learning in physics
- develop skills of independent working.

Throughout the course it is intended that you should develop an understanding of the impact of physics applications on everyday life, as well as the knowledge and skills that will allow you to reflect critically on scientific and media reports. You will learn to make your own judgements on the many issues that affect our society.

The course assignment helps develop investigative and communication skills. The course as a whole also helps develop literacy and numeracy skills.

The units

There are four units in the Higher Physics course. These are:

1. Our Dynamic Universe
2. Particles and Waves
3. Electricity
4. Researching Physics

What is covered in each unit?

The first three units cover a knowledge and understanding of physics, with each unit split into 'key areas' as follows:

Our Dynamic Universe (full unit)

- Motion – equations and graphs
- Forces, energy and power
- Collisions, explosions and impulse
- Gravitation
- Special relativity
- The expanding universe

Particles and Waves (full unit)

- The Standard Model
- Forces on charged particles
- Nuclear reactions
- Wave-particle duality
- Interference and diffraction
- Refraction of light
- Spectra

Electricity (half unit)

- Monitoring and measuring a.c.
- Current, potential difference, power and resistance
- Electrical sources and internal resistance
- Capacitors
- Conductors, semiconductors and insulators
- p-n junctions

A more detailed breakdown of the content of each key area is given in the 'Course Assessment Specification', which can be found in the Higher Physics section of the SQA website. All candidates should familiarise themselves with this information.

Researching Physics (half unit)

The Researching Physics unit (half unit) deals with the development of skills required to research an aspect of physics that is related to a key area in one of the other units of the course.

Assessment and the examination

Unit assessment

For each of the units (Our Dynamic Universe, Particles and Waves, and Electricity), there are two assessment outcomes.

Outcome 1

This outcome requires you to demonstrate that you can carry out an experimental procedure. There are six assessment standards within this outcome. These are:

1.1 Planning the experiment: Your plan needs to include:

- a clear statement of the aim of the experiment
- a hypothesis stating what you think might happen
- a statement describing the dependent and independent variables
- the variables that need to be kept constant
- the measurements/observations to be made
- the equipment/materials to be used
- a clear and detailed description of how the experiment should be carried out, including safety considerations.

1.2 Following procedures safely

1.3 Making and recording observations/measurements accurately

1.4 Presenting results in an appropriate format

1.5 Drawing valid conclusions

1.6 Evaluating experimental procedures

Outcome 2

This outcome requires you to demonstrate knowledge and understanding of the key areas of each unit, and apply scientific skills by making accurate statements and solving problems.

2.1 Making accurate statements: You will need to make accurate statements relating to each of the key areas defined in the course.

2.2 Solving problems: You will need to demonstrate problem-solving skills by:

- making predictions
- selecting information
- processing information
- analysing information.

Researching Physics

This unit covers some of the key skills necessary to undertake research in physics. The assessment standards for the two outcomes for the Researching Physics unit are given below.

Outcome 1

Apply skills of scientific inquiry, and draw on knowledge and understanding to research the underlying physics of a chosen topic by:

1.1 Gathering and recording information from two sources relating to the chosen topic.

Outcome 2

Apply skills of scientific inquiry to investigate, through experimentation, the underlying physics of a chosen topic by:

2.1 Planning/designing the practical investigation, including safety measures.

2.2 Carrying out the practical investigation safely, recording detailed observations/ measurements correctly.

Assignment

Once you have completed the Researching Physics unit, you are required to communicate your findings in an assignment.

The assignment will have 20 marks in total and is sent to the SQA to be externally marked. It contributes nearly 20% towards your final grade.

This assignment requires you to apply skills, knowledge and understanding to the investigation of a relevant topic in physics. The topic should draw on one or more of the key areas of the course:

- applying physics knowledge to new situations, interpreting information and solving problems
- selecting information and presenting information appropriately in a variety of forms
- processing information (using calculations, significant figures and units, where appropriate)
- drawing valid conclusions and giving explanations supported by evidence/justification
- communicating findings/information effectively.

The majority of the marks will be awarded for applying scientific inquiry and analytical thinking skills. The other marks will be awarded for applying knowledge and understanding related to the topic chosen.

More information on both the assignment and the Researching Physics unit can be found in Chapter 6.

The exam

The national examination lasts for 2 hours and 30 minutes, and has two sections equalling a total of 130 marks:

Section 1:	Twenty multiple choice questions	20 marks
Section 2:	Structured questions requiring written answers	135 marks (scaled to 100 marks)

Marks are distributed evenly across the units. This means that the two full units Our Dynamic Universe, and Particles and Waves will have a larger proportion of the overall marks than the half unit on Electricity.

The majority of marks are awarded for applying knowledge and understanding. The remaining marks are for applying scientific inquiry, scientific analytical thinking and problem-solving skills.

A data booklet containing relevant data and formulae will also be provided.

Skills, knowledge and understanding

This section gives a broad overview of the mandatory subject skills, knowledge and understanding that will be assessed in the course.

Processing information

Some standard units are not known well enough
You must learn the units of all the quantities in the Higher Physics course. It is useful to go through your notes or textbook and make your own list.

Lack of detail in answers to 'show that' questions
When asked to 'show that' a value is true, the first line should always quote a formula. This should be followed by substituting appropriate values and giving the final expected answer (the final answer must be identical to the value given in the question, for example, if you are asked to 'show that the distance is 2·2 m' and the answer on your calculator is 2·22 – the final line must be 2·2 m **not** 2·22 m). Missing out any of these stages means that a marker cannot be sure whether you have worked 'backwards' from the given answer.

Rounding the answers to intermediate calculations, leading to inaccuracies in the final answer
Some questions require you to carry out two consecutive calculations. The answer to the first calculation is used in the second calculation. Rounding the first answer can cause inaccuracy in the answer to the second calculation.

Candidates should not use an ellipsis in intermediate working. For example, where the answer to the first calculation is 2·1457, this should not be written as 2·14... . If rounding at this stage, you should use at least one more significant figure than should appear in the final answer.

Exam example 1

Q (a) A ray of red light of frequency 4.80×10^{14} Hz is incident on a glass lens as shown.

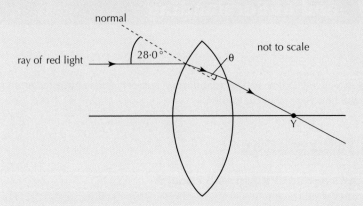

The ray passes through point Y after leaving the lens.

The refractive index of the glass is 1·61 for this red light.

(i) Calculate the wavelength of this light inside the lens.

A Correct answer:

$$\lambda_{air} = \frac{v_{air}}{f} = \frac{3.00 \times 10^8}{4.80 \times 10^{14}} = 6.25 \times 10^{-7} \ m$$

$$\lambda_{glass} = \frac{\lambda_{air}}{n} = \frac{6.25 \times 10^{-7}}{1.61} = 3.88 \times 10^{-7} \ m$$

Incorrect answer:

$$\lambda_{air} = \frac{v_{air}}{f} = \frac{3.00 \times 10^8}{4.80 \times 10^{14}} = 6.25 \times 10^{-7} = 6 \times 10^{-7} \ m$$

$$\lambda_{glass} = \frac{\lambda_{air}}{n} = \frac{6 \times 10^{-7}}{1.61} = 3.73 \times 10^{-7} \ m$$

The final answer here is inaccurate because the wavelength in air was inappropriately rounded to two significant figures **less** than the least number of significant figures given in the data.

Too many or too few significant figures given in final answer

If you give too many or too few significant figures in your final answer, it will result in a marking penalty. Each question has a range of acceptable significant figures, depending on the lowest number of significant figures given in the question. You are allowed two more and one less significant figure than this number.

Exam example 2

Q

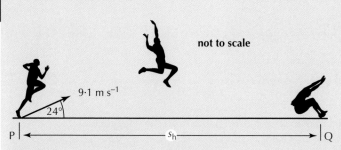

not to scale

9·1 m s⁻¹

24°

P ◄──────────────── s_h ──────────────► Q

An athlete takes part in a long jump competition. The athlete takes off from point P with an initial velocity of 9·1 m s⁻¹ at an angle of 24° to the horizontal and lands at point Q.

(a) Calculate:
    (i) the vertical component of the initial velocity of the athlete.

A $u_v = 9{\cdot}1 \sin 24°$
$u_v = 3{\cdot}7\ m\ s^{-1}$

The lowest number of significant figures given in the question is 2. This means the significant figure range for the final answer is 1–4. Acceptable answers are therefore 4, 3·7, 3·70 and 3·701.

Incorrect rounding of final answers

For example, a figure of '26·57', rounded to three significant figures should be '26·6' not '26·5'.

Example

Suppose that the data in a particular question is given to three significant figures. Your calculator displays the answer to the calculation as $4{\cdot}738709677 \times 10^{14}$. The third significant figure in this answer is the '3'. However, because the fourth figure is '8', the '3' is rounded up and the answer should be written as $4{\cdot}74 \times 10^{14}$. Do not round up when the next figure is less than '5'.

Poor knowledge of prefixes

The prefixes for the Higher Physics course are:

Prefix name	Prefix symbol	Power of ten
pico	p	$\times 10^{-12}$
nano	n	$\times 10^{-9}$
micro	μ	$\times 10^{-6}$
milli	m	$\times 10^{-3}$
kilo	k	$\times 10^{3}$
mega	M	$\times 10^{6}$
giga	G	$\times 10^{9}$
tera	T	$\times 10^{12}$

These prefixes are not listed in the examination paper or in the data sheet.

Uncertainties

Inability to change from absolute uncertainty to percentage uncertainty and vice versa

Example

A student plans to drop a ball from a height. She measures the height as 2·50 metres. She estimates that the uncertainty in this measurement is ±0·02 m – this means that she believes the value of the height lies between 2·48 and 2·52 metres.

She should record her result as 2·50 ±0·02 *m*.

The *absolute* uncertainty in her measurement is the value of the imprecision: ±0·02 *m*.

The *fractional* uncertainty in her measurement is the absolute uncertainty divided by the value of the measurement: $\dfrac{0·02}{2·50} = 0·008$.

The *percentage* uncertainty in her measurement is the fractional uncertainty multiplied by 100: $0·008 \times 100 = 0·8\%$.

Poor ability to express the final numerical result of an experiment in the form: final value ± uncertainty

For Higher Physics, you need to know that the best estimate of the percentage uncertainty in a final answer is equal to the larger (or largest) percentage uncertainty in the measurements used to calculate that answer. You must also be able to calculate an absolute uncertainty from a percentage uncertainty.

Example

Q A student is given the following measurements:

Current $= 0.025 \pm 0.001$ A

Voltage $= 12.0 \pm 0.25$ V

The student is asked to calculate the resistance and give its *absolute* uncertainty.

A Answer:

Resistance, $R = \dfrac{V}{I} = \dfrac{12}{0.025} = 480 \, \Omega.$

The absolute uncertainty in the current is ± 0.001 A.

The fractional uncertainty in the current is $\dfrac{0.001}{0.025} = 0.04$.

The percentage uncertainty in the current is 4% (Fractional uncertainty $\times 100$).

The absolute uncertainty in the voltage is ± 0.25 V.

The fractional uncertainty in the voltage is $\dfrac{0.25}{12} = 0.0208$.

The percentage uncertainty in the voltage is 2%.

The uncertainty in the resistance is therefore $\pm 4\%$ (the larger value).

4% of $480 = 480 \times \dfrac{4}{100} = 19.2$, which rounds to 20.

The resistance is therefore 480 ±20 Ω.

Graphs

Failure to label the origin of graphs with a zero

Many graphs in Physics are used to indicate a relationship between quantities, for example, that the charge stored in a capacitor is directly proportional to the potential difference across it.

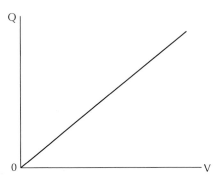

A graph can prove such a relationship when it is a straight, diagonal line passing through the origin. The point where the axes meet must therefore be shown as the point where both the quantities have their zero value. The usual way to do this is to write a zero ('0') at this point. Failure to do this can result in a loss of marks.

Failure to fully label the axes on graphs

The axes of all graphs, whether drawn on graph paper or sketched, should have each axis labelled with both the name of the quantity and its units. Any relevant values should also be marked on the axis where requested.

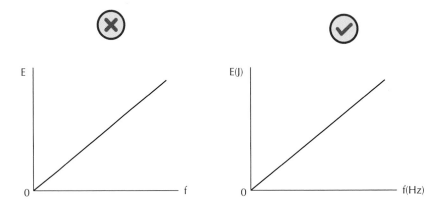

Sketch graphs too 'rough'

Make sure you use a ruler to draw the axes and the graph line (if it is straight). Take care to show whether your graph line intersects an axis (for example, in initial charging current with a capacitor) or whether it does not touch the axis.

Draw dotted reference lines to indicate any important values.

Avoid 'forcing' the line through the origin on graphs

When drawing a line of best fit, do not 'force' the line through the origin. Technically, if the straight line is drawn from data then the line should not project beyond the first and last data point, unless a relationship is requested.

Incorrect reading of scales of graphs in questions

Take your time to work out the value of one division along each axis. Double check your answer by making sure that you agree with the numbers given further along the axes.

When selecting data points from a graph, choose those that have 'definite' values, rather than those that are difficult to identify accurately.

Be careful of prefixes in units when selecting values from graphs, for example, time in milliseconds (ms) or current in milliamps (mA). Failure to include these can result in calculating the area under the graph, or gradient of the graph, incorrectly.

Knowledge and understanding

Knowledge and understanding questions are about being able to recall facts, symbols, diagrams, ideas and techniques.

Exam example 3

Q A student makes the following statements about an electron:

   I An electron is a boson.
  II An electron is a lepton.
 III An electron is a fermion.

Which of these statements is/are correct?

A I only
B II only
C III only
D I and II only
E II and III only

To answer this question, you need to be able to recall the definitions of 'fermions', 'leptons' and 'bosons':

Fermions is the collective name for all the matter particles (quarks and leptons).

Leptons are the electron, muon and tau along with their neutrinos.

Bosons are the force-mediating particles.

Therefore, statements II and III are correct.

A The answer is E.

Exam example 4

Q Which of the following proves that light is transmitted as waves?

A Light has a high velocity.
B Light can be reflected.
C Light irradiance reduces with distance.
D Light can be refracted.
E Light can produce interference patterns.

To answer this question, you need to know that the proof of wave motion is that an interference pattern can be produced.

A The answer is E.

Exam example 5

Q (*a*) Use band theory to explain how electrical conduction takes place in a pure semiconductor such as silicon.

Your explanation should include the terms: *electrons*, *valence band* and *conduction band*.

To answer this question, you need to have learned the electrical properties of conductors, insulators and semiconductors by using the electron population of the conduction and valence bands, and the energy difference between these bands.

A For a pure semiconductor:

Most of the electrons are in the valence band.

The band gap is small.

Electrons are thermally excited to the conduction band.

Charge flows when electrons are in the conduction band.

You may include a labelled diagram in your answer but this must be accompanied by a suitable explanation.

Exam example 6

Q A supply of e.m.f. 10·0 V and internal resistance r is connected in a circuit as shown in Figure 1.

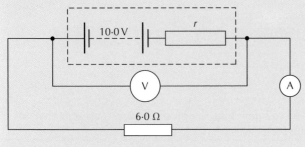

Figure 1

The meters display the following readings:

Reading on ammeter = 1·25 A

Reading on voltmeter = 7·50 V

(i) What is meant by an *e.m.f. of 10·0 V*?

To answer this question, you need to have learned the definition of 'e.m.f.'.

A The e.m.f. is the energy given to each coulomb of charge passing through the battery/power supply. So, in this case, there are **10 joules** of energy given to each coulomb of charge passing through the battery.

Exam example 7

Q (a) State what is meant by the term *capacitance*.

To answer this question, you need to have learned the definition of 'capacitance'.

It is useful when giving definitions of quantities to use an equation to help you derive the correct description. In this case $C = \dfrac{Q}{V}$, where C is capacitance; Q is charge; and V is potential difference.

A The definition of 'capacitance' is therefore the *charge stored per unit of voltage*.

Unfamiliar equations

A small number of marks in the exam paper may be awarded for working with unfamiliar equations. These are formulae that do not appear in the relationships sheet. They are also likely to be equations that are entirely new to you. It is important in these situations to read the question carefully, as all symbols and quantities will be clearly defined.

Exam example 8

Q Astronomers use the following relationship to determine the distance, d, to a star:

$$b = \frac{L}{4\pi d^2}$$

For a particular star the following data is recorded:

apparent brightness, $b = 4\cdot4 \times 10^{-10}$ W m^{-2}

luminosity, $L = 6\cdot1 \times 10^{30}$ W

Based on this information, the distance to this star is:

A $3\cdot3 \times 10^{19}$ m
B $1\cdot5 \times 10^{21}$ m
C $3\cdot7 \times 10^{36}$ m
D $1\cdot1 \times 10^{39}$ m
E $3\cdot9 \times 10^{39}$ m.

A
$$b = \frac{L}{4\pi d^2}$$

$$4\cdot4 \times 10^{-10} = \frac{6\cdot1 \times 10^{30}}{4\pi d^2}$$

$$4\cdot4 \times 10^{-10} \times 4\pi \times d^2 = 6\cdot1 \times 10^{30}$$

$$5\cdot529 \times 10^{-9} \times d^2 = 6\cdot1 \times 10^{30}$$

$$d^2 = 1\cdot0885 \times 10^{39}$$

$$d = 3\cdot3 \times 10^{19}\ m$$

The answer is A.

If you forget to take the square root of 'd', you will give the incorrect answer of D. This is a reminder that all of the options given in multiple-choice questions are possible answers, but only one of them uses the correct calculation.

Open questions

There will be 6 marks in the examination allocated to 'open questions' (two questions, each worth 3 marks). These are questions that do not have a specific answer but instead have a response that is **open-ended**. This means that there are many different ways in which the question can be answered.

This type of question will include the phrase 'using your knowledge of physics' so it is important that your answer includes a description of the **physics** concepts that relate to the question.

For **3 marks**, you must demonstrate a **good** understanding of the physics involved. This type of response might include a statement of the principle(s) involved, a relationship or equation, and the application of these to the question.

For **2 marks**, you must demonstrate a **reasonable** understanding of the physics involved. This type of response might make some statement(s) that are relevant to the question, including, for example, a statement of relevant principle(s) or identification of a relevant relationship or equation.

For **1 mark**, you must demonstrate a **limited** understanding of the physics involved. This type of response includes some statement(s) that are relevant to the question.

Exam example 9

Q A science textbook contains the following diagram of an atom:

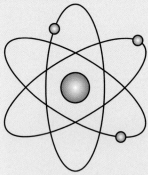

Use your knowledge of physics to comment on this diagram.

It is helpful before structuring your answer to identify the key physics concepts that you will include in your response. The model of the atom shown here is the **planetary** or **solar system model**. This is a largely inaccurate model as it does not represent a number of ideas about atoms.

The main ideas that can be explored here are:

- the Bohr model
- energy levels
- subatomic particles
- the standard model
- wave-particle duality
- the quantum model

However, you do not have to mention all of these in your response. For example, one idea could be explored in great detail to obtain 3 marks. Whichever concepts you choose should be explained and then related back to the question.

An example of this might be:

A The model shown is inaccurate as it does not represent the protons and neutrons that are within the nucleus of the atom. It also does not show up and down quarks which are the fundamental particles that make up protons and neutrons. Charges are not shown.

Common areas of difficulty in units 1, 2 and 3

Every year, the SQA publishes a report on the performance of candidates in each national examination. Amongst other things, these reports identify reasons why marks are lost and give advice on how candidates could better prepare for the examination in the future.

The advice given in this chapter has been constructed using the issues raised in these course reports. The majority of the rest of this book concentrates on techniques to help you perform well in the areas that candidates commonly find difficult.

Common areas of difficulty in

Our Dynamic Universe

Motion – equations and graphs

Poor understanding of the meaning of the term 'acceleration'

'Acceleration' means _how much the velocity changes each second_. By itself, the value of the acceleration tells you nothing about the speed or the distance travelled.

Exam example 10

Q An object has a constant acceleration of 3 m s^{-2}. This means that the:

A distance travelled by the object increases by 3 metres every second
B displacement of the object increases by 3 metres every second
C speed of the object is 3 m s^{-1} every second
D velocity of the object is 3 m s^{-1} every second
E velocity of the object increases by 3 m s^{-1} every second.

A The answer is E, simply because this is the definition of what 'acceleration' means.

Incorrect substitution of initial velocity, *u*, and final velocity, *v*

You need to remember that '*u*' represents initial velocity, and '*v*' represents final velocity. One method of remembering this is that 'initial' comes before 'final' and '*u*' comes before '*v*' in the alphabet.

The values must be substituted for the appropriate letters; it does not matter if one value is larger, or if the values are positive or negative. For example:

Exam example 11

Q To test the braking system of cars, a test track is set up as shown.

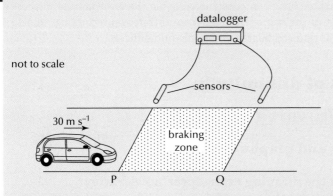

The sensors are connected to a datalogger, which records the speed of a car at both P and Q.

A car is driven at a constant speed of 30 m s⁻¹ until it reaches the start of the braking zone at P. The brakes are then applied.

(a) In one test, the datalogger records the speed at P as 30 m s⁻¹ and the speed at Q as 12 m s⁻¹. The car slows down at a constant rate of 9.0 m s⁻² between P and Q.

Calculate the length of the braking zone.

A $u = 30\ m\ s^{-1}$; $v = 12\ m\ s^{-1}$;

$a = -9 \cdot 0\ m\ s^{-2}$ (*car is slowing down so value is negative, indicating a deceleration*)

$v^2 = u^2 + 2as$

$12^2 = 30^2 + (2 \times (-9 \cdot 0) \times s)$

$144 = 900 - 18s$

$s = 42\ m$

If the values of u and v are substituted incorrectly, a maximum of 1 mark is awarded for the formula.

Failing to substitute a negative value of acceleration when an object is slowing down

Acceleration is a vector quantity and must be given a negative value if either a positive velocity is decreasing (as in the previous example) or the unbalanced force is in the opposite direction to a positive initial velocity.

Inconsistent positive and negative signs for initial velocity, u, final velocity, v, and acceleration, a

In a similar way to the previous two issues, you need to remember that 'u', 'v' and 'a' are all vector quantities. This means that their directions must be taken into account when substituting their values into an equation of motion.

You need to decide which direction you are going to consider as positive (up, down, right or left) and then be consistent in substituting any value as negative when it acts in the opposite direction. There is no 'correct' sign convention as long as one direction is taken as negative and the opposite direction as positive.

Exam example 12

Q A helicopter is **descending** vertically at a constant speed of 3·0 m s^{-1}. A sandbag is released from the helicopter. The sandbag hits the ground 5·0 s later.

What was the height of the helicopter above the ground at the time the sandbag was released?

A 15·0 m
B 49·0 m
C 107·5 m
D 122·5 m
E 137·5 m

A Taking *down* as positive,

$u = 3 \cdot 0 \ m \ s^{-1}; t = 5 \cdot 0 \ s; a = 9 \cdot 8 \ m \ s^{-2}; s = ?$

$s = ut + \dfrac{1}{2}at^2$

$s = \left(3 \cdot 0 \times 5 \cdot 0\right) + \left(\dfrac{1}{2} \times 9 \cdot 8 \times 5^2\right)$

$s = 15 + 122.5$

$s = 137 \cdot 5 \ m$

OR

Taking *up* as positive,

$u = -3 \cdot 0 \ m \ s^{-1}; t = 5 \cdot 0 \ s; a = -9 \cdot 8 \ m \ s^{-2}; s = ?$

$s = ut + \dfrac{1}{2}at^2$

$s = \left(-3 \cdot 0 \times 5 \cdot 0\right) + \left(\dfrac{1}{2} \times -9 \cdot 8 \times 5^2\right)$

$s = -15 + (-122.5)$

$s = -137 \cdot 5 \ m$

The negative sign indicates the downward displacement of sandbag is 137·5 m from helicopter to the ground.

The answer is E.

Drawing an acceleration-time graph from a velocity-time graph

The value of the acceleration should be calculated for each straight section of the velocity-time graph using $a = \dfrac{v - u}{t}$. The calculated values are then drawn as _horizontal_ lines on an acceleration-time graph because acceleration is constant during each period of time.

Exam example 13

Q (b) As the crate is moving up the slope, the rope snaps.

The graph shows how the velocity of the crate changes from the moment the rope snaps.

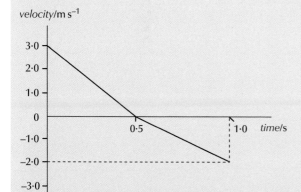

Copy the axes shown below and sketch the graph to show the acceleration of the crate between 0 and 1·0 s.

Appropriate numerical values are also required on the acceleration axis.

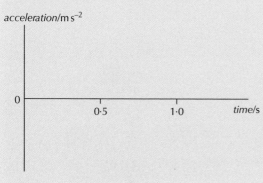

A From $t = 0\ s$ to $t = 0.5\ s$,

$$a = \frac{v - u}{t}$$

$$a = \frac{0 - 3}{0.5}$$

$$a = -6.0\ m\ s^{-2}$$

Note: Many candidates make the mistake of thinking that the acceleration changes from zero to $-6.0\ m\ s^{-2}$ during the first $0.5\ s$ but it remains constant at this value from $t = 0\ s$ to $t = 0.5\ s$.

From $t = 0.5\ s$ to $t = 1.0\ s$,

$$a = \frac{v - u}{t}$$

$$a = \frac{(-2.0 - 0)}{0.5}$$

$$a = -4.0\ m\ s^{-2}$$

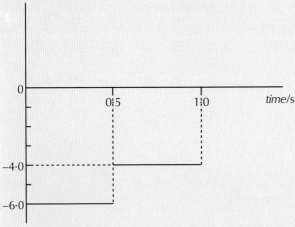

acceleration/m s^{-2}

From $t = 0\ s$ to $t = 0.5\ s$,

$$a = \frac{(v - u)}{t}$$

$$a = \frac{(0 - 3)}{0.5}$$

$$a = -6.0\ m\ s^{-2}$$

Poor understanding of the relationship between acceleration and velocity, especially in descriptive answers

The definition of 'acceleration' is the 'change in velocity each second'. This means that two different objects can have completely different speeds (or velocities) but have the same value of acceleration because *their velocities are changing at the same rate*.

For example, object A, which increases speed from $2 \cdot 5$ m s^{-1} to $7 \cdot 5$ m s^{-1} in one second, has the same acceleration as object B, which increases speed from $312 \cdot 5$ m s^{-1} to $317 \cdot 5$ m s^{-1} in one second.

The velocity of both objects has changed by $5 \cdot 0$ m s^{-1} in the same time interval.

Therefore, the acceleration of object A and of object B is $5 \cdot 0$ m s^{-2}.

Exam example 14

Q *(b)* As the crate is moving up the slope, the rope snaps.
The graph shows how the velocity of the crate changes from the moment the rope snaps.

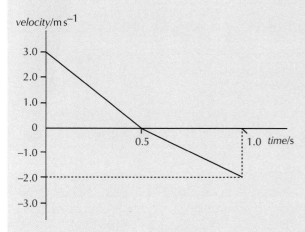

(i) Describe the motion of the crate during the first $0 \cdot 5$ s after the rope snaps.

A During the first $0 \cdot 5$ s, the velocity decreases from $3 \cdot 0$ m s^{-1} to 0 m s^{-1} at a constant rate.

This means that the object has a *constant negative acceleration* (or *constant deceleration*).

Forces, energy and power

Inability to state or derive the formula for the component of weight parallel/perpendicular to a slope

You need to know that these formulae are **not listed in the Physics relationships sheet**. This means that in the examination you need to be able to write them down from memory or derive them. Most candidates find it easier to memorise the formulae.

When an object of mass, m is on a slope at an angle of θ to the horizontal, the component of weight parallel to the slope is $mg \sin \theta$; the component of weight perpendicular to the slope is $mg \cos \theta$.

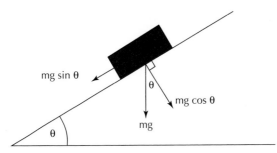

Exam example 15

Q A fairground ride consists of rafts which slide down a slope into water.

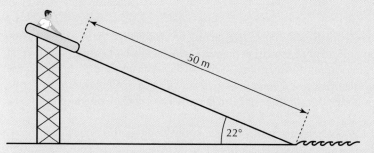

The slope is at an angle of 22° to the horizontal. Each raft has a mass of 8·0 kg. The length of the slope is 50 m.

A child of mass 52 kg sits in a raft at the top of the slope. The raft is released from rest. The child and raft slide together down the slope into the water. The force of friction between the raft and slope remains constant at 180 N.

(a) Calculate the component of weight, in newtons, of the child and raft down the slope.

A (a) *component of weight* $= mg \sin \theta$

$$= 60 \times 9·8 \times \sin 22°$$

$$= 220 \, N$$

The total mass is the mass of the raft (8 kg) plus the mass of the child (52 kg).

Poor ability to describe and explain how a number of forces combine to affect the acceleration or velocity of an object

To improve your ability to answer such questions, first of all you need to ensure you understand Newton's second law, $F = ma$.

In the formula, F represents the <u>unbalanced force</u> acting on an object of mass m. This means that you need to <u>combine all of the forces</u> acting on the object into <u>one resultant force</u>. It is essential to take into account the <u>directions</u> of the forces when combining them because each force is a <u>vector</u> quantity. It is a good idea to make a <u>sketch</u> of all of the forces acting on an object in a free-body diagram. This makes it far easier to calculate the resultant force.

The acceleration of the object is always in the <u>same direction</u> as the <u>unbalanced force</u>. As the <u>unbalanced force increases</u>, the <u>acceleration increases</u>, and as the acceleration increases, the velocity increases.

Exam example 16

Q A fairground ride consists of rafts which slide down a slope into water.

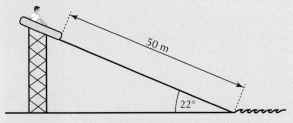

The slope is at an angle of 22° to the horizontal. Each raft has a mass of 8·0 kg. The length of the slope is 50 m.

A child of mass 52 kg sits in a raft at the top of the slope. The raft is released from rest. The child and raft slide together down the slope into the water. The force of friction between the raft and slope remains constant at 180 N.

(a) Calculate the component of weight, in newtons, of the child and raft down the slope.

(b) Show by calculation that the acceleration of the child and raft down the slope is 0·67 m s⁻².

(c) Calculate the speed of the child and raft at the bottom of the slope.

(d) A second child of smaller mass is released from rest in an identical raft at the same starting point. The force of friction is the same as before.

How does the speed of this child and raft at the bottom of the slope compare with the answer to part (c)?

Justify your answer.

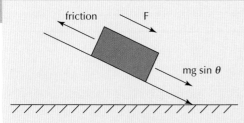

(a) *component of weight* $= mg \sin \theta$

$$= 60 \times 9{\cdot}8 \times \sin 22°$$

$$= 220 \, N$$

(b) $F = component\ of\ weight - friction$

$$= 220 - 180 = 40 \, N$$

$$a = \frac{F}{m} = \frac{40}{60} = 0{\cdot}67 \, m \, s^{-2}$$

(c) $u = 0 \, m \, s^{-1}$; $a = 0{\cdot}67 \, m \, s^{-2}$; $s = 50 \, m$; $v = ?$

$$v^2 = u^2 + 2as$$

$$v^2 = 0 + 2 \times 0{\cdot}67 \times 50$$

$$v^2 = 67$$

$$v = 8{\cdot}2 \, m \, s^{-1}$$

(d) Smaller mass means that the component of weight acting down the slope decreases. If the component of weight decreases and the frictional force remains constant then the unbalanced force also decreases. The acceleration decreases so the final velocity also decreases.

Note: Part (d) can also be answered via a calculation. Choose a smaller value of mass (less than 60 kg) and recalculate parts (a), (b) and (c).

Exam example 17

Q A car of mass 1200 kg pulls a horsebox of mass 700 kg along a straight, horizontal road. They have an acceleration of $2 \cdot 0$ m s^{-2}.

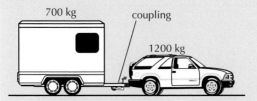

700 kg   coupling

1200 kg

Assuming that the frictional forces are negligible, the tension in the coupling between the car and the horsebox is:

A 500 N
B 700 N
C 1400 N
D 2400 N
E 3800 N.

A The tension in the coupling is the unbalanced force, which is acting on the horsebox and causing it to accelerate.

Apply Newton's second law to the *horsebox*. The unbalanced force causing it to accelerate at $2 \cdot 0$ *m s^{-2}* is:

$F = ma$

$F = 700 \times 2 \cdot 0$

$F = 1400$ *N (to the right)*.

The answer is C. (The mass of the car is irrelevant for this calculation.)

Exam example 18

Q Two boxes on a frictionless horizontal surface are joined together by a string. A constant horizontal force of 12 N is applied as shown.

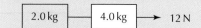

The tension in the string joining the two boxes is:

A 2·0 N
B 4·0 N
C 6·0 N
D 8·0 N
E 12 N.

A To answer this type of question, you need to apply Newton's second law *twice*. Firstly, apply it to the *whole system* to find the acceleration. Then, apply it again to a *part of the system* to find unbalanced force on that part.

For the whole system:

$F = ma$

$12 = 6·0 \times a$

$a = 2·0 \text{ m s}^{-2}$ (to the right)

The tension in the string provides the force to accelerate the 2·0 kg mass.

For the 2·0 kg mass:

$F = ma$

$F = 2·0 \times 2·0$

$F = 4·0 \text{ N}$ (to the right)

The answer is B.

Exam example 19

Q A skydiver of total mass 85 kg is falling vertically.

At one point during the fall, the air resistance on the skydiver is 135 N.

The acceleration of the skydiver at this point is:

A 0.6 m s^{-2}
B 1.6 m s^{-2}
C 6.2 m s^{-2}
D 8.2 m s^{-2}
E 13.8 m s^{-2}.

A $W = mg = 85 \times 9.8 = 833 \text{ N}$ *(downwards)*

The unbalanced force on the skydiver is:

Weight − Air resistance = 833 − 135 = 698 N (down)

$$a = \frac{F}{m} = \frac{698}{65} = 8.21 \ m \ s^{-2}$$

The answer is D.

135 N

833 N

Many candidates answer B for this question. You can only arrive at this incorrect solution if you neglect to calculate the *unbalanced* force before using F = ma.

Collisions, explosions and impulse

Lack of understanding of the vector nature of impulse and momentum and the link between these quantities

Impulse = average force × time

Force is a vector quantity and so requires a direction. This means that impulse is also a vector quantity.

The units for impulse are Newton seconds (N s).

Momentum = mass × velocity

Velocity is a vector quantity and so requires direction. This means that momentum is also a vector quantity.

The units for momentum are kilogram metres per second (kg m s^{-1}).

The relationship between momentum and impulse is:

Impulse = change in momentum

$Ft = mv - mu$

where $mv - mu$ = final momentum − initial momentum

Exam example 20

Q A golfer hits a ball of mass $5{\cdot}0 \times 10^{-2}$ kg with a golf club. The ball leaves the tee with a velocity of 80 m s^{-1}. The club is in contact with the ball for a time of 0·10 s.

The average force exerted by the club on the ball is:

A $6{\cdot}25 \times 10^{-4}$ N
B 0·025 N
C 0·4 N
D 4 N
E 40 N.

A $Ft = mv - mu$

$F \times 0{\cdot}1 = 50 \times 80 - 50 \times 0$

$0{\cdot}1F = 4{\cdot}0$

$F = 40\ N$

The answer is E.

Exam example 21

Q Beads of liquid moving at high speed are used to move threads in modern weaving machines.

(a) In one design of machine, beads of water are accelerated by jets of air as shown in the diagram:

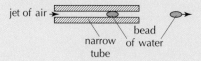

jet of air

narrow tube

bead of water

Each bead has a mass of 2.5×10^{-5} kg.

When designing the machine, it was estimated that each bead of water would start from rest and experience a constant unbalanced force of 0·5 N for a time of 3·0 ms.

(i) Calculate:

    (A) the impulse on a bead of water

    (B) the speed of the bead as it emerges from the tube.

A (i) (A) *Impulse = average force × time*

$$= 0.5 \times 3.0 \times 10^{-3}$$

$$= 1.5 \times 10^{-3} \, N\,s$$

(B) $Ft = mv - mu$

$$1.5 \times 10^{-3} = 2.5 \times 10^{-5} \, v - 2.5 \times 10^{-5} \times 0$$

$$v = \frac{1.5 \times 10^{-3}}{2.5 \times 10^{-5}}$$

$$v = 60 \, m\,s^{-1}$$

Top tip

When performing calculations with change in momentum, you need to be careful when substituting values for v and u. Velocity is a vector, so its direction must be taken into account, meaning that if u and v are in opposite directions then one of the values will be negative.

Exam example 22

Q The apparatus shown below is used to test concrete pipes.

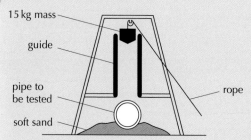

When the rope is released, the 15 kg mass is dropped and falls freely through a distance of 2·0 m on to the pipe.

(a) In one test, the mass is dropped on to an uncovered pipe.

(i) Calculate the speed of the mass just before it hits the pipe.

(ii) When the 15 kg mass hits the pipe, the mass is brought to rest in a time of 0·02 s. Calculate the size and direction of the average unbalanced force on the **pipe**.

A (a) (i) Taking downwards as the positive direction:

$u = 0; a = 9·8; s = 2·0; v = ?$

$v^2 = u^2 + 2as$

$v^2 = 0^2 + 2 \times 9·8 \times 2·0$

$v^2 = 39·2$

$v = 6·26 \, m \, s^{-1}$

The mass is travelling downwards at 6·26 $m \, s^{-1}$ just before it hits the pipe.

(ii) Taking downwards as the positive direction:

When the mass collides with the pipe, its initial velocity, u, is 6·26 $m \, s^{-1}$. After 0·02 s, it comes to a stop, so its final velocity, v, is 0 $m \, s^{-1}$.

$Ft = mv - mu$

$F \times 0{\cdot}02 = 15 \times 0 - 15 \times 6{\cdot}26$

$0{\cdot}02\, F = -93{\cdot}9$

$F = \dfrac{-93{\cdot}9}{0{\cdot}02}$

$F = -4695 = -4700\ N$

The negative value for this answer shows that *the force on the mass is upwards.*

Candidates who substitute the values for u and v incorrectly will get a final answer of +4700 N. However, this is regarded as wrong physics at the substitution stage and most of the marks are lost.

Finally, to answer the question, you need to apply Newton's third law, which states that 'for every action force, there is an equal and opposite reaction force'. Therefore, the pipe experiences an equal and opposite force due to this collision and so the average unbalanced force on the pipe is *4700 N downwards.*

Top tip

There are often questions using force-time graphs. Impulse is equal to the area under a force-time graph. This means that you can calculate this area to give the change in momentum of an object when a force is applied to it.

Exam example 23

Q The graph shows the force which acts on an object over a time interval of 8 seconds.

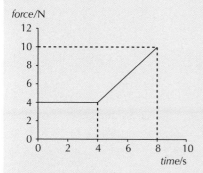

The momentum gained by the object during this 8 seconds is:

A 12 kg m s^{-1}
B 32 kg m s^{-1}
C 44 kg m s^{-1}
D 52 kg m s^{-1}
E 72 kg m s^{-1}.

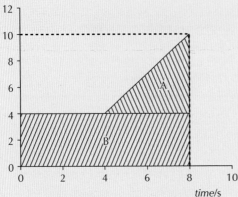

The gain (or change) in momentum is equal to the impulse.

So:

Impulse = area under force-time graph

Impulse = area of A + area of B

$$Impulse = \left(\frac{1}{2} \times 4 \times 6\right) + \left(4 \times 8\right)$$

Impulse = 12 + 32

Impulse = 44 *N s*

OR 44 kg m s^{-1}

The answer is C.

Some candidates fail to calculate the *full* area under the graph. For example, only calculating the area of triangle A and neglecting to include the area of rectangle B.

Note that the units of impulse can be used to express change in momentum and vice versa. This is because impulse = change in momentum.

Top tip

When carrying out calculations with force-time graphs, time intervals are often small. Be careful to convert units such as milliseconds to the standard unit of time (seconds).

Incorrect use of positive and negative signs for velocities in calculations
The law of conservation of linear momentum is used to work out what happens in collisions and explosions. You must be very careful to decide which direction you wish to be positive (for example, left to right). Anything moving in the opposite direction must then be given a *negative* velocity (and momentum).

Exam example 24

Q Two trolleys travel towards each other in a straight line as shown.

The trolleys collide. After the collision the trolleys move as shown below.

What is the speed v of the 2·0 kg trolley after the collision?

A 1·25 m s^{-1}
B 1·75 m s^{-1}
C 2·0 m s^{-1}
D 4·0 m s^{-1}
E 5·0 m s^{-1}

A Taking left to right as the positive direction,

Total momentum before the collision = Total momentum after the collision

$$m_1u_1 + m_2u_2 = m_1v_1 + m_2v_2$$

$$(6{\cdot}0 \times 2{\cdot}0) + (2{\cdot}0 \times (-1{\cdot}0)) = (6{\cdot}0 \times 1{\cdot}0) + (2{\cdot}0 \times v)$$

$$12 - 2 = 6 + 2v$$

$$2v = 4$$

$$v = 2\ m\ s^{-1}$$

The answer is C.

Note: Notice the negative sign for the 1·0 *m s*$^{-1}$ to the left.

Poor understanding of the meaning of an 'elastic' collision

Many candidates make the mistake of thinking that an elastic collision means that two objects do not stick together. However, the correct definition of an elastic collision is 'a collision where the total kinetic energy is conserved'.

The following question involves most of the issues mentioned in this section about impulse and momentum.

Exam example 25

Q Two ice skaters are initially skating together, each with a velocity of 2·2 m s^{-1} to the right as shown.

2.2 m s^{-1}

skater R

skater S

The mass of skater R is 54 kg. The mass of skater S is 38 kg.

Skater R now pushes skater S with an average force of 130 N for a short time. This force is in the same direction as their original velocity.

As a result, the velocity of skater S increases to 4·6 m s^{-1} to the right.

4.6 m s^{-1}

skater R   skater S

(a) Calculate the magnitude of the change in momentum of skater S.
(b) How long does skater R exert the force on skater S?
(c) Calculate the velocity of skater R immediately after pushing skater S.
(d) Is this interaction between the skaters elastic?
    You must justify your answer by calculation.

A   (a)   Change in momentum $= mv - mu$

$$= (38 \times 4 \cdot 6) - (38 \times 2 \cdot 2)$$

$$= 174 \cdot 8 - 83 \cdot 6$$

$$= 91 \cdot 2 \; kg \, m \, s^{-1}$$

(b) $Ft = mv - mu$

$130 \times t = 91 \cdot 2$ **Note:** Notice there is no need to substitute values for m, v and u as change in momentum has already been calculated

$$t = \frac{91 \cdot 2}{130}$$

$$t = 0 \cdot 7 \; s$$

(c) $m_1 u_1 + m_2 u_2 = m_1 v_1 + m_2 v_2$

$(54 \times 2 \cdot 2) + (38 \times 2 \cdot 2) = (54 \times v) + (38 \times 4 \cdot 6)$

$54v = 118 \cdot 8 + 83 \cdot 6 - 174 \cdot 8$

$54v = 27 \cdot 6$

$v = 0 \cdot 51 \; m \, s^{-1}$

(d)   Total kinetic energy before collision:

$$\frac{1}{2} m_1 u_1^{\,2} + \frac{1}{2} m_2 u_2^{\,2}$$

$$= \frac{1}{2} \times 54 \times 2 \cdot 2^2 + \frac{1}{2} \times 38 \times 2 \cdot 2^2$$

$$= 223 \; J$$

Total kinetic energy after collision:

$$\frac{1}{2} m_1 v_1^{\,2} + \frac{1}{2} m_2 v_2^{\,2}$$

$$= \frac{1}{2} \times 54 \times 0 \cdot 51^2 + \frac{1}{2} \times 38 \times 4 \cdot 6^2$$

$$= 409 \; J$$

Total kinetic energy is not constant and so this collision is *not an elastic collision* (it would now be called an inelastic collision).

Gravitation

Incorrect use of Newton's Universal Law of Gravitation to solve problems involving force, masses and their distance of separation

Newton's law of gravitation states that the gravitational attraction that exists between two objects is directly proportional to the mass of each object and inversely proportional to the square of their distance apart.

$$F = \frac{Gm_1m_2}{r^2}$$

G is the universal gravitational constant and has the value $6.67 \times 10^{-11}\,\mathrm{m^3\,kg^{-1}\,s^{-2}}$.

m_1 and m_2 are the two masses measured in kilograms.

The distance, r, between the two objects is the distance between their **centres of mass**. A common mistake made is to measure from the surfaces of the objects (planets, moons etc.).

Difficulty in linking Newton's Law of Gravitation with gravitational field strength

Every mass creates a gravitational field around itself. A gravitational field is a region where other masses will experience a gravitational force. Gravitational field strength is the force on a 1 kg mass placed in the field.

As g is the force per unit mass, it is related to the mass, M, of the object providing the force by the following expression:

$$g = \frac{F}{m} = \frac{GMm}{mr^2} = \frac{GM}{r^2}$$

Like gravitational force, beyond the surface of the object, the value of g follows an inverse square law.

Exam example 26

Q A space probe of mass $5\cdot60 \times 10^3$ kg is in orbit at a height of $3\cdot70 \times 10^6$ m above the surface of Mars.

space probe

Mars

not to scale

The mass of Mars is $6\cdot42 \times 10^{23}$ kg.

The radius of Mars is $3\cdot39 \times 10^6$ m.

(a) Calculate the gravitational force between the probe and Mars.

(b) Calculate the gravitational field strength of Mars at this height.

A (a) $F = \dfrac{G\, m_1 m_2}{r^2}$

Note: The distance between the two objects (r) includes the radius of Mars *and* the orbit height of the space probe. Many candidates used the radius or height of Mars *only*.

$$F = \frac{6\cdot67 \times 10^{-11} \times 6\cdot42 \times 10^{23} \times 5\cdot60 \times 10^3}{\left(3\cdot39 \times 10^6 + 3\cdot70 \times 10^6\right)^2}$$

$F = 4\cdot77 \times 10^3\, N$

(b) This is simply an application of $W = mg$, using the force of the probe from part (a) and the mass:

$$g = \frac{W}{m}$$

$$g = \frac{4770}{5600}$$

$g = 0\cdot852\, N\, kg^{-1}.$

Special relativity

Poor understanding of the postulates of special relativity

Einstein built his special theory of relativity using two basic ideas or postulates:

- The laws of physics are the same for all observers in inertial frames of reference.
- The speed of light (in a vacuum) is the same for all observers.

This means that observers in all frames of reference will measure the speed of light in a vacuum to be $3.00 \times 10^8 \ m \ s^{-1}$.

Inability to define frames of reference when describing measurements of space and time

Measurements of space and time for a moving observer are changed relative to a stationary observer.

This means you need to be careful to define from which frame of reference the measurements are being made. In other words, you must clearly state which observer will experience time dilation or length contraction.

Exam example 27

Q *(b)* On the plane, the students discuss the possibility of travelling at relativistic speeds.

    (i) The students consider the plane travelling at $0.8 \ c$ relative to a stationary observer. The plane emits a beam of light towards the observer.

    State the speed of the emitted light as measured by the observer.

    Justify your answer.

    (iii) One of the students states that the clocks on board the plane will run slower when the plane is travelling at relativistic speeds.

    Explain whether or not this statement is correct.

A *(b)* (i) $3.00 \times 10^8 \ m \ s^{-1}$

    The speed of light is the same for all observers.

    (iii) The statement is correct. From the **stationary observer's frame of reference**, the measured time is slower.

    Note: The explanation must include a statement that refers to or implies a frame of reference.

Difficulty in using appropriate relationships to solve problems involving length contraction and time dilation

One consequence of the speed of light being the same for all observers is that the time experienced by different observers is not always the same. A process that takes a certain time to occur in a moving frame of reference is observed to take a **longer** time from the point of view of someone in a different frame of reference.

$$t' = \frac{t}{\sqrt{1-\left(\dfrac{v^2}{c^2}\right)}}$$

t = time interval in a frame of reference

t' = time interval measured by an observer in a **different** frame of reference

v = the relative velocity of the two frames of reference

c = the speed of light in a vacuum

Another consequence of the speed of light being the same for all observers is the shortening or contraction of length when an object is moving. An object moving at extremely high speed will appear shorter than an identical object that is stationary, when both are observed by the same stationary observer. That is, the length of a rocket travelling at 1×10^8 m s^{-1} will appear shorter in length compared to an identical rocket stationary on Earth when both are observed by the same stationary observer on Earth.

$$l' = l\sqrt{1-\left(\frac{v^2}{c^2}\right)}$$

l = distance in a frame of reference

l' = distance measured by an observer in a **different** frame of reference

v = the relative velocity of the two frames of reference

c = the speed of light in a vacuum

Exam example 28

Q Muons are sub-atomic particles produced when cosmic rays enter the atmosphere about 10 km above the surface of the Earth.

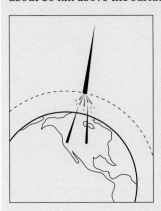

Muons have a mean lifetime of $2 \cdot 2 \times 10^{-6}$ s in their frame of reference.
Muons are travelling at $0 \cdot 995\, c$ relative to an observer on Earth.

(b) Calculate the mean lifetime of the muons as measured by the observer on Earth.

A The observer on Earth will measure the time as being longer, therefore t = 2.2 × 10⁻⁶ s.
Many candidates confuse t and t'. t' should always be the *longer* time.

$$t' = \frac{t}{\sqrt{1 - \left(\dfrac{v^2}{c^2}\right)}}$$

$$t' = \frac{2 \cdot 2 \times 10^{-6}}{\sqrt{1 - \left(\dfrac{0 \cdot 995^2}{1^2}\right)}}$$

$$t' = 2 \cdot 2 \times 10^{-5}\ s$$

Exam example 29

Q *(b)* On the plane, the students discuss the possibility of travelling at relativistic speeds.

(ii) According to the manufacturer, the length of the plane is 71 m.

Calculate the length of the plane travelling at 0·8 c as measured by the stationary observer.

A

$$l' = l\sqrt{1 - \left(\frac{v^2}{c^2}\right)}$$

$$l' = 71\sqrt{1 - 0 \cdot 8^2}$$

$$l' = 43 \, m$$

l' is the contracted length, in other words the shorter length. Therefore l' should always be less than l.

The Expanding Universe

Incorrect use of Doppler effect calculations

For the source moving **towards** the observer:

$$f_{observed} = f_{source}\left(\frac{v}{v - v_{source}}\right)$$

For the source moving **away** from the observer:

$$f_{observed} = f_{source}\left(\frac{v}{v + v_{source}}\right)$$

Top tip

v = speed of the wave (sound or light).

The equation given in the relationships sheet is:

$$f_o = f_s\left(\frac{v}{v \pm v_s}\right).$$

So you need to remember whether to use the + or −. It is a good idea to check that your answers give the expected increase or decrease in frequency. The frequency of the sound or light increases as the source moves towards the observer and decreases as the source moves away from the observer.

Exam example 30

Q The siren on an ambulance is emitting sound with a constant frequency of 900 Hz. The ambulance is travelling at a constant speed of 25 m s^{-1} as it approaches and passes a stationary observer. The speed of sound in air is 340 m s^{-1}.

Which row in the table shows the frequency of the sound heard by the observer as the ambulance approaches and as it moves away from the observer?

	Frequency as ambulance approaches (Hz)	Frequency as ambulance moves away (Hz)
A	900	900
B	971	838
C	838	900
D	971	900
E	838	971

A **Note:** Write the equation from the data sheet first.

$$f_o = f_s\left(\frac{v}{v \pm v_s}\right)$$

Note: When the ambulance is moving towards the observer, the observed frequency f_o increases. You therefore want to multiply f_s by a larger number. This means that you will use a '−' instead of a '+'.

$$f_o = 900\left(\frac{340}{340-25}\right)$$

$$f_o = 971\ Hz$$

Note: When the ambulance is moving away from the observer, the observed frequency f_o decreases. You therefore want to multiply f_s by a smaller number. This means that you will use a '+' instead of a '−'.

$$f_o = 900\left(\frac{340}{340+25}\right)$$

$$f_o = 838\ Hz$$

The answer is B.

Poor awareness of evidence supporting the Big Bang theory

If the Big Bang did happen then there should be some radiation left over that can be detected in the present day. The electromagnetic radiation produced at the Big Bang would be redshifted due to the expansion of the universe. This radiation has been detected and is the same in all directions. It is called the cosmic microwave background.

The most abundant elements in the universe are hydrogen (70–75%) and helium (30–25%). These amounts of hydrogen and helium cannot be explained purely by nuclear fusion processes in stars. However, it can be accounted for by the formation of lighter elements shortly after the Big Bang.

Exam example 31

Q (a) Experimental work at CERN has been described as "*recreating the conditions that occurred just after the Big Bang*".

Describe what scientists mean by the *Big Bang theory* and give **one** piece of evidence which supports this theory.

A The universe was initially in a hot and very dense state and then rapidly expanded.

One piece of evidence that supports this theory is the cosmic microwave background.

Note: Alternatively, you could use the abundance of hydrogen/helium.

Common areas of difficulty in Particles and Waves

The Standard Model

Comparing the mass of objects in terms of orders of magnitude

Physics deals with quantities across a vast range of orders of magnitude, from the tiniest of scales studied in particle physics to the largest of scales used when studying distances in space. We express the order of magnitude of objects in powers of 10.

For example, an object that is 1000 times larger is three orders of magnitude bigger.

Exam example 32

Q (b) In July 2012, scientists at CERN announced that they had found a particle that behaved in the way that they expected the Higgs boson to behave. Within a year, this particle was confirmed to be a Higgs boson.

This Higgs boson had a mass-energy equivalence of 126 GeV ($1 \text{ eV} = 1 \cdot 6 \times 10^{-19}$ J).

(i) Show that the mass of the Higgs boson is $2 \cdot 2 \times 10^{-25}$ kg.

(ii) Compare the mass of the Higgs boson with the mass of a proton in terms of orders of magnitude.

A (b) (i) $126 \text{ } GeV = 126 \times 10^9 \times 1 \cdot 6 \times 10^{-19}$ **Note:** The electronvolt is a unit of energy.

$= 2 \times 10^{-8} J$

$E = mc^2$

$2 \times 10^{-8} = m \times (3 \times 10^8)^2$

$m = 2 \cdot 2 \times 10^{-25} kg$

(ii) **Note:** The mass of a proton can be found in the Higher Physics data sheet.

Mass of proton $= 1 \cdot 673 \times 10^{-27} kg$

Mass of Higgs boson $= 2 \cdot 2 \times 10^{-25} kg$

Note: Next, divide the mass of the Higgs boson by the mass of the proton:

$2 \cdot 2 \times 10^{-25} \div (1 \cdot 673 \times 10^{-27}) = 130$

The Higgs boson is approximately 100 times bigger than the proton. But because we express orders of magnitude in powers of 10: The Higgs boson is *2 orders of magnitude bigger.*

Forces on charged particles

Inability to define the term 'potential difference'
The potential difference between two points is a measure of the work done in moving one coulomb of charge between the two points. If one joule of work is done in moving one coulomb of charge between two points, the potential difference between the two points is one volt.

Not only do these definitions need to be remembered but you also need to be able to adapt them to include appropriate values for a particular question.

Exam example 33

Q The diagram below shows the basic features of a proton accelerator. It is enclosed in an evacuated container.

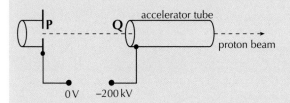

Protons released from the proton source start from rest at **P**.
A potential difference of 200 kV is maintained between **P** and **Q**.

(a) What is meant by the term *potential difference of 200 kV*?

A Each coulomb of positive charge gains 200,000 J of kinetic energy as it moves from P to Q.

Confusion between the two formulas for calculating energy, 'QV' and '$\frac{1}{2}$QV' (see also page 89)
When a charge, Q, is moved through a potential difference of 'V' volts, the work done, 'W', is calculated from W = QV.

Exam example 34

Q A potential difference of 5000 V is applied between two metal plates. The plates are 0·10 m apart. A charge of +2·0 mC is released from rest at the positively charged plate as shown:

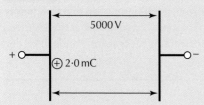

The kinetic energy of the charge just before it hits the negative plate is:

A $4·0 \times 10^{-7}$ J
B $2·0 \times 10^{-4}$ J
C 5·0 J
D 10 J
E 500 J.

A *Gain in kinetic energy = work done on charge Q = QV*

$= 2·0 \times 10^{-3} \times 5000$

$= 10$ J

The answer is D.

Nuclear reactions

Rounding values of masses before the loss of mass is calculated

In both nuclear fission and nuclear fusion, there is a very small loss in mass. This 'lost' mass becomes energy according to Einstein's equation $E = mc^2$. However, the loss in mass is so small that significant inaccuracies are introduced if figures are rounded before calculating its value. For this reason, in questions on fission and fusion, masses are often given to several significant figures. Do not round answers in these calculations until you obtain the final value for the energy released.

Q (a) The following statement represents a fusion reaction:

$$4\,{}^{1}_{1}\text{H} \rightarrow {}^{4}_{2}\text{He} + 2\,{}^{0}_{1}\text{e}$$

The masses of the particles involved in the reaction are shown in the table.

Particle	Mass (kg)
${}^{1}_{1}\text{H}$	$1{\cdot}673 \times 10^{-27}$
${}^{4}_{2}\text{He}$	$6{\cdot}646 \times 10^{-27}$
${}^{0}_{1}\text{e}$	negligible

(i) Calculate the energy released in this reaction.

A $\Delta m = 4 \times 1{\cdot}673 \times 10^{-27} - 6{\cdot}646 \times 10^{-27}$

$\Delta m = 4{\cdot}6 \times 10^{-29} \, kg$

$E = mc^2$

$E = 4{\cdot}6 \times 10^{-29} \times (3{\cdot}00 \times 10^{8})^2$

$E = 4{\cdot}14 \times 10^{-12} \, J$

Poor awareness of coolant and containment issues in nuclear fusion reactors

A controlled fusion reaction can be achieved in a device known as a Tokamak. Hydrogen molecules are brought to extremely high temperatures of over 100 million Kelvin. At these temperatures, the hydrogen gas molecules separate into atoms and then into electrons and nuclei, forming a plasma (the fourth state of matter).

At such high temperatures, the ordinary solid matter of a container would melt or evaporate. Because the plasma consists of moving charged particles, strong magnetic fields can be used to confine the plasma and keep it moving in circles in the shape of a hollow doughnut or torus. The magnetic field also keeps the plasma away from the inner surfaces of the container. If the plasma was to touch the surfaces, it would melt the container.

Exam example 35

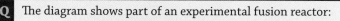

Q The diagram shows part of an experimental fusion reactor:

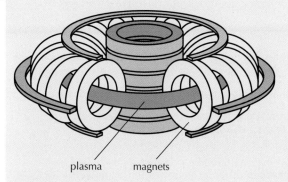

plasma   magnets

(c) Magnetic fields are used to contain the plasma inside the fusion reactor. Explain why it is necessary to use a magnetic field to contain the plasma.

A Plasma would melt the sides of the reactor.

OR

Plasma could damage or destroy the container.

Note: Many candidates failed to identify any containment issues in a fusion reactor.

Wave particle duality

Poor understanding of the variables and processes involved in the photoelectric effect

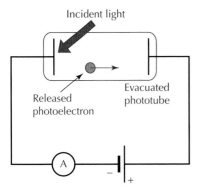

The main points to be known and understood about the photoelectric effect are:

- in some circumstances, 'light' can cause a charged metal to discharge
- 'light' can mean infrared, visible light, ultraviolet radiation or any of the other members of the electromagnetic spectrum
- the 'light' consists of a stream of photons (bundles of electromagnetic radiation)
- one photon collides with one electron and gives that electron all its energy
- the energy carried by each photon is given by $E = hf$ (where 'f' is frequency and 'h' is Planck's constant)
- the minimum energy an electron needs in order to escape from the metal is called 'the work function'
- an electron is released only if the photon's energy is equal to, or greater than, the work function
- when the photon's energy is greater than the work function, the excess energy becomes kinetic energy of the electron
- increasing the irradiance of the 'light' increases the number of photons per second
- increasing the irradiance of the 'light' increases the number of electrons released per second (provided that the energy of the photon is greater than or equal to the work function)
- increasing the irradiance of the 'light' does not increase the maximum kinetic energy of each photoelectron.

Exam example 36

Q Ultraviolet radiation is incident on a clean zinc plate. Photoelectrons are ejected.

The clean zinc plate is replaced by a different metal which has a lower work function. The same irradiance of ultraviolet radiation is incident on this metal.

Compared to the zinc plate, which of the following statements is/are true for the new metal?

  I The maximum speed of the photoelectrons is greater.
 II The maximum kinetic energy of the photoelectrons is greater.
III There are more photoelectrons ejected per second.

A I only
B II only
C III only
D I and II only
E I, II and III

A The source of the radiation has not changed, so from $E = hf$ photons have the same energy as before. This means that the electrons receive the same energy as before.

A lower work function means that electrons need less energy to escape from the surface of the metal. This means the electrons have greater kinetic energy than before. Statement II is correct.

Greater kinetic energy means greater speed ($E_K = \frac{1}{2}mv^2$). Statement I is correct.

The same irradiance means there are the same number of photons per second. Therefore, the same number of electrons are released per second. Statement III is incorrect.

The answer is D.

Confusion between the terms 'threshold frequency' and 'work function'

The work function of a metal is the minimum energy an electron needs to receive in order to escape from the surface of a metal. It is measured in joules (J).

The threshold frequency, f_o, is the minimum frequency of 'light' required to cause the emission of photoelectrons (photoemission). It is measured in hertz (Hz).

The relationship between these two quantities is:

work function $= hf_o$

Inability to explain the effect of increasing irradiance on both the photoelectric current and the maximum kinetic energy of photoelectrons

Increasing the irradiance of the 'light' increases the number of photons per second. This increases the number of photoelectrons released per second and so increases the photoelectric current.

However, an electron only receives the energy of one photon. As long as the same frequency of light is used, each photon still has the same energy as before ($E = hf$) and so there is no change in the kinetic energy of an emitted electron.

Exam example 37

Q When light of frequency f is shone on to a certain metal, photoelectrons are ejected with a maximum velocity v and kinetic energy E_k.

Light of the same frequency but twice the irradiance is shone on to the same surface. Which of the following statements is/are correct?

  I Twice as many electrons are ejected per second.
 II The speed of the fastest electron is 2 v.
III The kinetic energy of the fastest electron is now 2 E_k.

A I only
B II only
C III only
D I and II only
E I, II and III

A Twice the irradiance means that there are double the number of photons per second. This doubles the number of electrons released per second. Statement I is correct.

Using the light of the same frequency means that each photon has the same energy as before. This means that each electron receives the same energy as before. Each electron therefore has the same kinetic energy and same speed as before. Statements II and III are incorrect.

The answer A.

Interference and diffraction

Incomplete descriptions of what happens to cause constructive interference and destructive interference

'Interference' is a wave effect caused by two or more sets of waves overlapping. In your answer, you must say that the waves *meet*.

Constructive interference occurs when waves meet *in phase* with each other. Alternatively, you can describe constructive interference by saying 'crests meet crests and troughs meet troughs'.

Destructive interference occurs when waves meet perfectly out of phase with each other. This can also be described as 'crests meet troughs'.

In order to see a good interference effect, the waves used should be **coherent**.

Coherent means that the waves from the sources have a constant phase difference.

Poor understanding of the relationship between path difference and interference patterns

Path difference is a measurement of distance. If two wave sources meet at a point and one wave has travelled further than the other, then there is a path difference between the two waves. In other words, it is the difference in the distances travelled by the wave sources to the point where interference occurs. It is found by subtracting the smaller distance from the larger distance.

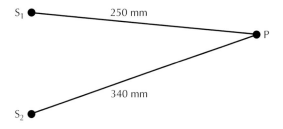

Path difference to $P = 340 - 250 = 90$ mm

If this path difference is a whole number of wavelengths then a maxima will be produced. If the path difference is a whole number plus a half wavelength (as at point P in the diagram below), then a minima will be produced.

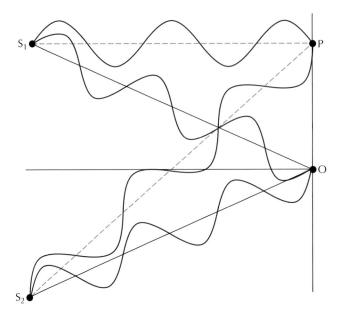

For constructive interference:

path difference = $m\lambda$ where m = 0, 1, 2 ....

For destructive interference:

path difference = $\left(m + \dfrac{1}{2}\lambda \right)$ where m = 0, 1, 2 ....

Exam example 38

Q (a) An experiment with microwaves is set up as shown below.

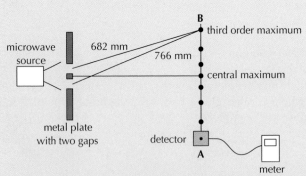

(ii) The measurements of the distance from each gap to a third-order maximum are shown. Calculate the wavelength of the microwaves.

A Path difference = 766 − 682 = 84 mm

Third-order maximum means m = 3

$3\lambda = 84\ mm$

$\lambda = \dfrac{84}{3}$

$\lambda = 28\ mm$

Poor understanding of the relationship involving grating spacing, wavelength, order number and angle to the maximum

A diffraction grating is a series of narrow, parallel slits usually etched into glass. Using more slits lets more light through and produces a series of bright sharp lines.

The path difference is equivalent to $d\sin\theta$, where d is the separation of the slits and θ is the angle of deviation. If the path difference is an integral multiple of one wavelength then a maxima will be produced.

Therefore:

$d\sin\theta = m\lambda$

The first maximum will occur when m = 0; that is, when there is no path difference. This is referred to as the zero-order maximum. When m = 1, the path difference is exactly one wavelength. This is called the first-order maximum.

Light of a single wavelength or frequency has a single colour and is called monochromatic light. When monochromatic light is passed through a diffraction grating, a series of bright fringes of one colour (such as red) will be produced.

Since the order of the maximum and subsequently, the angle of deviation (θ), depend on the wavelength of the light then different colours of light will be produced at different angles. This means a diffraction grating can be used to disperse white light into the colours of the visible spectrum. At m = 0, where there is no path difference, a central white maximum will be produced. The first-order maximum will show a spectrum of colour with violet light (the shortest wavelength) at the smallest angle and red light (the longest wavelength) at the largest angle.

Top tip

The lines on a diffraction grating can often be given in 'lines per mm' or 'lines per metre'. You need to be able to calculate 'd', the spacing between slits, from data given in this form. To do this, convert the value to lines per metre then use:

$d = \dfrac{1}{no.\ of\ lines/m}.$

Exam example 39

Q *(b)* In the second experiment, a ray of white light is incident on a grating.

not to scale

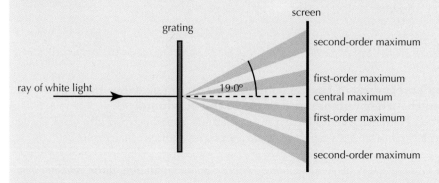

The angle between the central maximum and the second-order maximum for red light is 19·0°.

The frequency of this red light is $4·57 \times 10^{14}$ Hz.

(i) Calculate the distance between the slits on this grating.

(ii) Explain why the angle to the second-order maximum for blue light is different to that for red light.

A (i) **Note:** Calculate wavelength first, remembering that the speed of light in air is $3·00 \times 10^8 \, m \, s^{-1}$.

$v = f\lambda$

$3·00 \times 10^8 = 4·57 \times 10^{14} \times \lambda$

$\lambda = 656·5 \times 10^{-9} \, m$

Note: Next substitute values into the diffraction grating formula. The question mentions 'the second-order maximum' so therefore m = 2.

$d \sin \theta = m\lambda$

$d \sin 19·0° = 2 \times 656·5 \times 10^{-9}$

$d = 4·03 \times 10^{-6} \, m$

(ii) **Note:** The explanation can be generated from the diffraction grating formula.

From $d \sin \theta = m\lambda$

Different colours have different wavelengths.

The values of 'm' and 'd' are the same.

Therefore, the angle of deviation θ must be different for different wavelengths.

Refraction of light

Poor definitions of refraction

When a wave moves from one medium to another, the wave changes speed, for example, a light wave moving from air into glass. The change in speed can lead to a change in direction. The change in direction when a wave moves from one medium to another is called refraction.

Poor ability in completing diagrams to show the path taken by a ray of light as it passes from air into and through another medium

There are two key elements to answering questions like this successfully:

1. knowledge of the law of refraction and experience of using it
2. knowledge and experience of using simple geometry

The law of refraction is given in the Higher relationships sheet as $n = \dfrac{\sin\theta_1}{\sin\theta_2}$

where the angle for the less dense material (usually air) is on the top line of the equation.

Angles are measured between the ray of light and the normal line. Remember that the normal line makes an angle of $90°$ with the material's surface.

In terms of geometry, you should know that the angles in any triangle add up to $180°$. Alternate (or 'z') angles between parallel lines are equal.

Exam example 40

Q A physics student investigates what happens when monochromatic light passes through a glass prism or a grating.

(a) The apparatus for the first experiment is shown below.

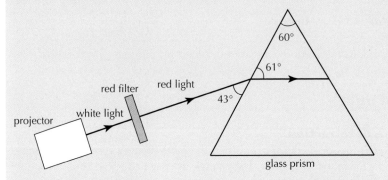

(i) Calculate the refractive index of the glass for the red light.

(ii) Sketch a diagram which shows the ray of red light before, during and after passing through the prism. Mark on your diagram the values of all relevant angles.

A (i) **Note:** Notice that the angles given in the diagram are not between the ray and the normal. The normal line is also not shown on the diagram so will need to be drawn in.

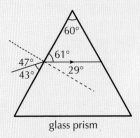

glass prism

$$n = \frac{\sin\theta_1}{\sin\theta_2}$$

$$n = \frac{\sin 47°}{\sin 29°}$$

$$n = 1.51$$

(ii) **Note:** Firstly, use geometry to find the angle of incidence at the right-hand face (31°). Then use the law of refraction for a second time to calculate the angle of refraction in air as the light emerges from the prism. Medium '1' is air and medium '2' is glass.

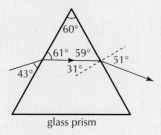

glass prism

$$n = \frac{\sin\theta_1}{\sin\theta_2}$$

$$1\cdot51 = \frac{\sin\theta_1}{\sin 31°}$$

$$\sin\theta_1 = 1\cdot51 \times \sin 31°$$

$$\theta_1 = 51°$$

Lack of knowledge of the link between the refractive index of a medium and the frequency of the light travelling through the medium

The different colours of the visible spectrum have different frequencies. Light at the red end of the spectrum has a lower frequency than light at the violet end of the spectrum. This means that a glass prism has a lower value of refractive index for red light than, for example, blue light. It also means that a ray of blue light will be refracted by a greater amount than a ray of red light passing through the prism along the same initial path.

Hence, the refractive index depends on the frequency of the incident light.

Exam example 41

Q A ray of red light of frequency 4.80×10^{14} Hz is incident on a glass lens as shown.

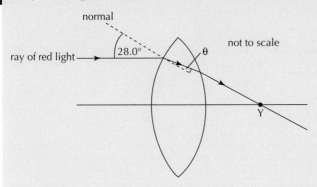

(b) The ray of red light is now replaced by a ray of blue light.

The ray is incident on the lens at the same point as above.

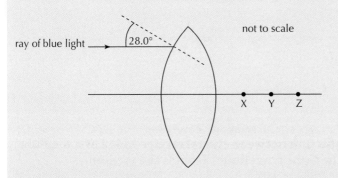

Through which point, X, Y or Z, will this ray pass after leaving the lens?
You must justify your answer.

A Refractive index depends on the frequency of the light. The glass therefore has a greater refractive index for blue light than for red light. As a result, the ray of blue light is refracted more than the ray of red light. It therefore passes through point X:

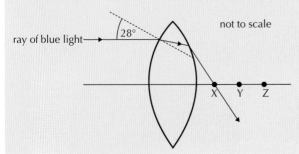

Poor explanations of what is meant by the critical angle, θ_c

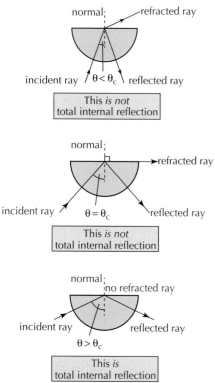

The critical angle θ_c is shown in the second diagram. It is the angle at which the refracted ray just emerges along the face of the D-shaped prism. The angle of refraction is 90°.

Many candidates wrongly state that the critical angle is the angle at which total internal reflection occurs. In fact, total internal reflection only occurs when the angle of incidence is *greater than* the critical angle.

Spectra

Poor explanations of how absorption lines in the spectrum of sunlight are produced

A heated mixture of many elements, such as the Sun, appears to produce a continuous spectrum (a spectrum with no visible separate lines). Early in the nineteenth century, Josef von Fraunhofer discovered that the spectrum of the Sun was crossed with a large number of dark lines.

It was shown that the wavelengths associated with these lines corresponded to the emission spectra of known elements. Light emitted from hotter regions of the Sun produces a continuous spectrum, but this light travels through cooler regions in the upper atmosphere. As it does so, atoms in these cooler regions absorb energy from the light at specific wavelengths to produce the absorption lines.

Fraunhofer had discovered a method of determining the elements present in the Sun's atmosphere.

Exam example 42

Q Light from the Sun is used to produce a visible spectrum.

A student views this spectrum and observes a number of dark lines as shown.

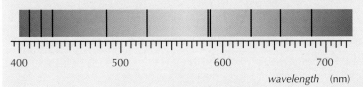

wavelength (nm)

(a) Explain how these dark lines in the spectrum of sunlight are produced.

A Particular wavelengths of light are absorbed *in the Sun's outer atmosphere*.

Note: Most candidates neglected to include that the wavelengths of light were absorbed in the Sun's atmosphere or the Sun's outer layers. 'The atmosphere' is considered too vague.

Can also replace 'particular wavelengths of light' with 'particular frequencies of light'.

Poor understanding of electron transitions in atoms and the absorption and emission of light

In an atom, electrons are arranged around the nucleus in specific orbits or energy levels. When an atom absorbs energy, electrons move upwards from lower energy levels to higher energy levels.

When electrons move down from higher energy levels to lower energy levels, the atom emits energy as a photon of 'light'. The larger the gap between these energy levels, the greater the energy of the emitted photon. This means that when electrons drop through a larger gap in energy levels, 'light' of a higher frequency and smaller wavelength is emitted.

Exam example 43

Q The diagram represents some electron transitions between energy levels in an atom.

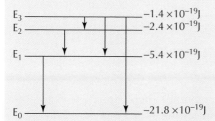

$$E_3 \quad\quad -1.4 \times 10^{-19} \text{J}$$
$$E_2 \quad\quad -2.4 \times 10^{-19} \text{J}$$
$$E_1 \quad\quad -5.4 \times 10^{-19} \text{J}$$
$$E_0 \quad\quad -21.8 \times 10^{-19} \text{J}$$

The radiation emitted with the shortest wavelength is produced by an electron making transition:

A E_1 to E_0
B E_2 to E_1
C E_3 to E_2
D E_3 to E_1
E E_3 to E_0.

A The shortest wavelength corresponds to the highest frequency.

From $E = hf$ the highest frequency means the greatest energy.

Greatest energy means a downward transition through the largest energy gap, which is E_3 to E_0.

The answer is E.

Top tip

The energy will be shown in energy level diagrams as a negative number. The negative value indicates that energy is required for the electron to move upwards and/or escape from the atom. Remember this when calculating the difference in energy (ΔE) between levels.

Common areas of difficulty in Electricity

Monitoring and measuring a.c.

Incorrect readings of the displays on an oscilloscope

Oscilloscopes are used to examine electrical signals. The y-axis displays the voltage and the x-axis displays the time. The y-gain setting controls the *height* of the displayed signal. This setting tells you the number of volts (or millivolts) for each division up and down the screen.

The amplitude of an alternating signal is the distance between the peak and the trough, divided by two as shown below:

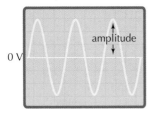

A y-gain setting of 5·0 *V/div* produces a smaller amplitude than one of 2·0 *V/div*. The peak voltage is calculated by multiplying the y-gain setting by the number of divisions of the amplitude. The r.m.s. voltage is then calculated using the formula $V_{peak} = \sqrt{2}\,V_{rms}$.

The timebase setting on an oscilloscope controls the *width* of a wave on the screen. This setting gives the number of seconds (or milliseconds) for each division across the screen. A setting of 5·0 ms/div produces double the number of waves across the screen than a setting of 2·5 ms/div.

The timebase setting is multiplied by the number of divisions across the screen for one wave (that is, one wavelength) to calculate the period, T, of the signal. The frequency, f, of the signal can then be calculated from $f = \dfrac{1}{T}$.

Exam example 44

Q An alternating voltage is applied to the input of an oscilloscope. The diagram below shows the trace displayed.

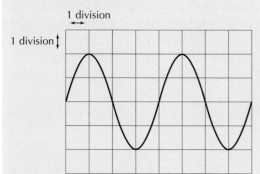

The Y-gain is set at 20 V/division. The timebase is set at 10 ms/division.

Identify the row in the table below that shows the peak voltage and the frequency of the signal.

	Peak voltage/V	Frequency/Hz
A	14·2	10
B	28	25
C	40	10
D	40	25
E	80	25

A The amplitude of the signal is 2 divisions.

Since the y-gain is set at 20 *V/div*,

Peak voltage $= 2 \times 20 = 40$ *V.*

Horizontally, one wave covers 4 divisions.

Since the timebase is set at 10 *ms/div*,

Period $= 4 \times 10 \ ms = 40 \ ms$

$$f = \frac{1}{T} = \frac{1}{40 \times 10^{-3}} = 25 \ \text{Hz.}$$

The answer is D.

Exam example 45

Q A signal from a power supply is displayed on an oscilloscope.

The trace on the oscilloscope is shown:

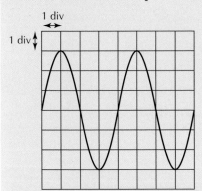

The timebase is set at 0·01 s/div and the Y-gain is set at 4·0 V/div.

Which row in the table shows the r.m.s. voltage and the frequency of the signal?

	r.m.s. voltage/V	frequency/Hz
A	8·5	25
B	12	25
C	24	25
D	8·5	50
E	12	50

A The amplitude of the signal is 3 divisions.

Since the y-gain is set at 4 *V/div*,

Peak voltage $= 3 \times 4 = 12\ V$

$$V_{rms} = \frac{V_{peak}}{\sqrt{2}} = \frac{12}{\sqrt{2}} = 8 \cdot 5\ V.$$

Horizontally, one wave covers 4 divisions.

Since the timebase is set at 0·01 *s/div*,

Period $= 4 \times 0 \cdot 01 = 0 \cdot 04\ s$

$$f = \frac{1}{T} = \frac{1}{0 \cdot 04} = 25\ Hz.$$

The answer is A.

Current, potential difference, power and resistance

Ohm's law (V = IR) applies in all circuits. It is often useful to apply this relationship to the whole circuit by using total resistance and supply voltage to calculate circuit current. You can then use V = IR again and apply it to a part of the circuit, using the circuit current and a resistance value to calculate the voltage across that resistor. Alternatively, use the potential divider formula given in the relationships sheet.

Exam example 46

Q A battery of e.m.f. 12 V and internal resistance 3·0 Ω is connected in a circuit as shown:

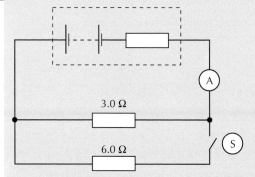

3.0 Ω

6.0 Ω

When switch **S** is closed, the ammeter reading changes from:

A 2·0 A to 1·0 A
B 2·0 A to 2·4 A
C 2·0 A to 10 A
D 4·0 A to 1·3 A
E 4·0 A to 6·0 A.

A One way of answering this question is as follows:

When switch S is open:

External resistance $= 3{\cdot}0\ \Omega$

Total circuit resistance $= 3{\cdot}0 + 3{\cdot}0 = 6{\cdot}0\ \Omega$

$$I = \frac{V_S}{R_T} = \frac{12}{6{\cdot}0} = 2{\cdot}0\ A$$

When S is closed:

The external resistance is the parallel combination of $3{\cdot}0\ \Omega$ and $6{\cdot}0\ \Omega$

$$\frac{1}{R_T} = \frac{1}{R_1} + \frac{1}{R_2}$$

$$\frac{1}{R_T} = \frac{1}{3} + \frac{1}{6}$$

$R_T = 2{\cdot}0\ \Omega$

Total circuit resistance $= 2{\cdot}0 + 3{\cdot}0 = 5{\cdot}0\ \Omega$

$$I = \frac{V_S}{R_T} = \frac{12}{5{\cdot}0} = 2{\cdot}4\ A$$

The answer is B.

Poor understanding of current and potential differences in circuits containing a number of resistors in series and parallel

Series circuits and parallel circuits 'behave' in different ways.

Series circuits

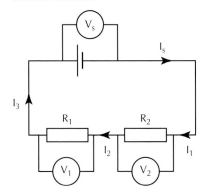

Current:

$$I_S = I_1 = I_2 = I_3$$

Voltage:

$$V_S = V_1 + V_2 + \cdots$$

Resistance:

$$R_T = R_1 + R_2 + \cdots$$

Parallel circuits

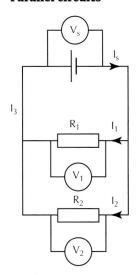

Current:

$$I_S = I_1 + I_2 + \cdots$$

Voltage:

$$V_S = V_1 = V_2 = \ldots$$

Resistance:

$$\frac{1}{R_T} = \frac{1}{R_1} + \frac{1}{R_2} + \cdots$$

Top tip

Only the circuit rules for resistance are given in the Higher relationships sheet. The rules for current and voltage must be remembered.

Exam example 47

Q The circuit below shows resistors connected as a potential divider.

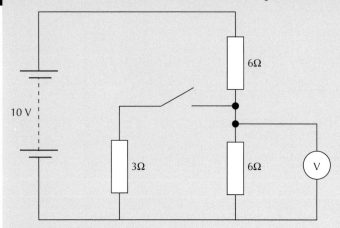

Calculate the reading on the voltmeter:

(a) When the switch is open;

(b) When the switch is closed.

A (a) **Note:** Calculate the total circuit resistance (no current in the 3Ω resistor when the switch is open):

$R_T = R_1 + R_2$

$R_T = 6{\cdot}0 + 6{\cdot}0$

$R_T = 12\ \Omega$

Note: Calculate the circuit current:

$I = \dfrac{V_S}{R_T} = \dfrac{10}{12} = 0{\cdot}83\ A$

Note: Apply this current to the $6{\cdot}0\ \Omega$ resistor:

$V = IR = 0{\cdot}83 \times 6{\cdot}0 = 5V$

(b) **Note:** Calculate the resistance of the parallel branch:

$\dfrac{1}{R_T} = \dfrac{1}{R_1} + \dfrac{1}{R_2}$

$$\frac{1}{R_T} = \frac{1}{6} + \frac{1}{3}$$

$$R_T = 2\,\Omega$$

Total circuit resistance $= 2{\cdot}0 + 6{\cdot}0 = 8{\cdot}0\ \Omega$

Note: Calculate the circuit current:

$$I = \frac{V_S}{R_T} = \frac{10}{8} = 1{\cdot}25\ A$$

Note: Apply this current to the parallel branch ($2{\cdot}0\ \Omega$):

$$V = IR = 1{\cdot}25 \times 2{\cdot}0 = 2{\cdot}5V$$

Note: The voltage is the same across both resistors connected in parallel, which is why we apply the current of $1{\cdot}25\ A$ to the total resistance of the parallel branch. The current will split between the parallel branches, so it is easier to treat this section of the circuit as one resistor rather than two.

Exam example 48

Q A battery has an e.m.f. of $6{\cdot}0$ V and internal resistance of $2{\cdot}0\ \Omega$.

(a) What is meant by an *e.m.f. of 6·0 V*?

A Each coulomb of charge gains 6 J of electrical potential energy as it passes through the battery/source.

Electrical sources and internal resistance

Inability to define the term e.m.f.

E.m.f. stands for *electro-motive force*. The e.m.f. of a source is the electrical potential energy supplied to each coulomb of charge which passes through the source.

You need to learn this well enough to be able to write it out from memory. You also need to be able to adapt it to include appropriate values for a particular question.

Poor understanding of the relationship involving e.m.f., terminal potential difference (t.p.d.), current and internal resistance

The effect of the internal resistance is to reduce the potential difference across the external circuit. When a very high resistance voltmeter is connected directly to a cell, the reading on the voltmeter is the cell's e.m.f. as no current is flowing.

When the cell is connected to a closed circuit, the voltmeter will read a value that is less than the e.m.f. The reading on the voltmeter is called the terminal potential difference (t.p.d.).

The voltage drop is caused by the internal resistance (r) of the cell and is called the 'lost' volts. Therefore:

e.m.f. = t.p.d. + lost volts

$E = V + Ir$

$E = IR + Ir$

$E = I(R + r)$

Exam example 49

Q A circuit is set up as shown:

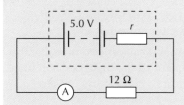

The e.m.f. of the battery is 5·0 V.

The reading on the ammeter is 0·35 A.

The internal resistance r of the battery is:

A 0·28 Ω
B 0·80 Ω
C 1·15 Ω
D 2·3 Ω
E 3·2 Ω.

A $e.m.f. = 5·0 \text{ V}; I = 0·35 \text{ A}; R = 12 \text{ Ω}; r = ?$

Note: 'R' is the external resistance and 'r' is the internal resistance.

$E = I(R + r)$

$5·0 = 0·35(12 + r)$

$5·0 = 0·35(12 + r)$

$5·0 = 4·2 + 0·35r$

Note: You can see at this point that t.p.d. is equal to 4.2 V and lost volts is 0.8 V.

$0·35r = 0·8$

$r = 2·3 \text{ Ω}$

The answer is D.

Capacitors

Poor understanding of the processes of charging and discharging a capacitor

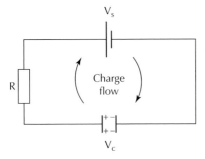

- Initially the capacitor is uncharged.
- Electrons pass from the negative terminal of the battery to one plate of the capacitor and away from the other plate.

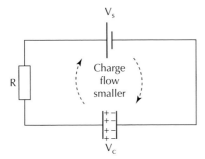

- Voltage across the capacitor (V_c) is initially very small.
- As charging proceeds, more negative charge is on one plate of the capacitor. The increased charge repels more negative charges arriving, hence current drawn becomes less. The increasing positive charge on the other plate of the capacitor makes removal of electrons from that plate even more difficult as time passes, due to electrostatic attraction.
- Potential difference across the capacitor increases as the build-up of charge on both plates increases. This also opposes the supply p.d., making the transfer of charge decrease with time.

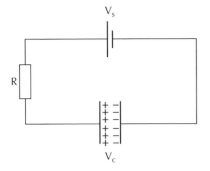

- Eventually, charge transfer decreases until there is no current. This is because the potential difference across the plates of the capacitor is equal to the potential difference across the battery ($V_s = V_c$).

When a capacitor is connected in series with a resistor, the time taken for a capacitor to charge is determined by the values of resistance and capacitance. Larger values of capacitance and larger values of resistance will increase the time it takes for a capacitor to charge or discharge.

Current-time graphs for differing values of capacitance and resistance are shown below for a charging capacitor.

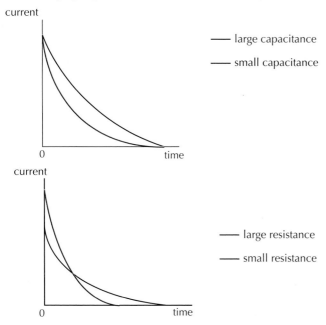

Note that the initial current (charging current) is smaller for a larger value of resistance and the capacitor takes longer to charge.

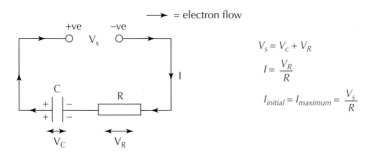

$$V_s = V_c + V_R$$

$$I = \frac{V_R}{R}$$

$$I_{initial} = I_{maximum} = \frac{V_s}{R}$$

Exam example 50

Q (a) State what is meant by the term *capacitance*.

(b) An uncharged capacitor, C, is connected in a circuit as shown.

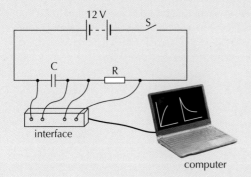

The 12 V battery has negligible internal resistance.

Switch S is closed and the capacitor begins to charge.

The interface measures the current in the circuit and the potential difference (p.d.) across the capacitor. These measurements are displayed as graphs on the computer.

Graph 1 shows the p.d. across the capacitor for the first 0·40 s of charging.

Graph 2 shows the current in the circuit for the first 0·40 s of charging.

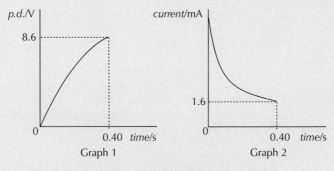

(i) Determine the p.d. **across resistor R** at 0·40 s.

(ii) Calculate the resistance of R.

(iii) The capacitor takes 2·2 seconds to charge fully.

At that time it stores 10·8 mJ of energy.

Calculate the capacitance of the capacitor.

A

(a) Capacitance = the charge stored per volt (this comes from the formula $C = \frac{Q}{V}$)

(b) (i) $V_R = V_S - V_C = 12 - 8 \cdot 6 = 3 \cdot 4$ V

(ii) $R = \frac{V_R}{I} = \frac{3 \cdot 4}{1 \cdot 6 \times 10^{-3}} = 2125\ \Omega$

(iii) When fully charged the p.d. across the capacitor = $V_S = 12$ V

$$E = \frac{1}{2}CV^2$$

$$10 \cdot 8 \times 10^{-3} = \frac{1}{2} \times C \times 12^2$$

$$10 \cdot 8 \times 10^{-3} = 72\ C$$

$$C = 1 \cdot 5 \times 10^{-4}\ F$$

Confusion between the two formulas for calculating energy, 'QV' and '$\frac{1}{2}$QV' (see also page 58)

When charge Q is transferred from one plate of a capacitor to the other plate, the potential difference between the plates is not constant, but instead increases from zero to the voltage of the supply. We therefore cannot use $W = QV$. The average potential difference during the charging process in a capacitor is '$\frac{1}{2}V$' and therefore the work done is '$\frac{1}{2}QV$'. **Note:** Only use '$\frac{1}{2}QV$' when dealing with capacitors; at all other times use QV.

Exam example 51

Q An uncharged 2200 µF capacitor is connected in a circuit as shown:

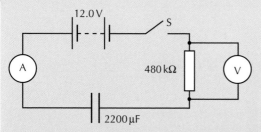

The battery has negligible internal resistance.

(c) Calculate the **maximum** energy the capacitor stores in this circuit.

A The maximum potential difference across the capacitor in this circuit is 12 V.

$$E = \frac{1}{2}CV^2$$

$$E = \frac{1}{2} \times 2200 \times 12^2$$

$$E = 0\cdot16\ J$$

Top tip

There are three equations given in the relationships sheet that can be used to calculate the energy stored in a capacitor:

$$E = \frac{1}{2}QV \qquad E = \frac{1}{2}\frac{Q^2}{C} \qquad E = \frac{1}{2}CV^2$$

These three equations appear side by side on the relationships sheet and should be used only for capacitance calculations. $W = QV$ is located on its own in a different column on the relationships sheet.

Poor ability to analyse current, potential differences and stored energy when a capacitor and resistor are connected in series with a d.c. supply

The sum of the potential differences across the capacitor and resistor is equal to the supply voltage.

$$V_R = V_S - V_C$$

The current in the circuit is not affected by capacitance and can be calculated using $I = \dfrac{V_R}{R}$

where V_R is the voltage across the resistor.

The energy stored in a capacitor depends on the potential difference, V_C, across the plates of the capacitor.

Exam example 52

Q A student investigates the charging and discharging of a 2200 μF capacitor using the circuit shown:

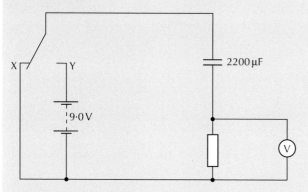

The 9·0 V battery has negligible internal resistance.

Initially, the capacitor is uncharged and the switch is at position X.

The switch is then moved to position Y and the capacitor charges fully in 1·5 s.

(a) (i) Sketch a graph of the p.d. across the **resistor** against time while the capacitor charges. Appropriate numerical values are required on both axes.

   (ii) The resistor is replaced with one of higher resistance.

      Explain how this affects the time taken to fully charge the capacitor.

   (iii) At one instant during the charging of the capacitor the reading on the voltmeter is 4·0 V.

      Calculate the charge stored by the capacitor at this instant.

(b) Using the same circuit in a later investigation the resistor has a resistance of 100 kΩ. The switch is in **position Y** and the capacitor is fully charged.

   (i) Calculate the maximum energy stored in the capacitor.

   (ii) The switch is moved to position X. Calculate the maximum current in the resistor.

A (a) (i) During the charging process, the potential difference *across the capacitor* increases from zero to 9·0 V. The p.d. across the resistor therefore decreases from 9·0 V to zero.

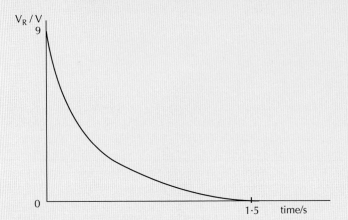

(ii) A higher resistance causes the current at any instant in time to be smaller than it was before. This means that it takes a *longer* time to transfer the same quantity of charge to fully charge the capacitor.

(iii) $V_R = 4{\cdot}0 \ V$

So $V_C = V_S - V_R = 9{\cdot}0 - 4{\cdot}0 = 5{\cdot}0 \ V$

$Q = CV = 2200 \times 10^{-6} \times 5{\cdot}0 = 0{\cdot}011 \ C$

(b) (i) When the capacitor is fully charged, $V_C = 9{\cdot}0 \ V$

$$E = \frac{1}{2}CV^2$$

$$E = \frac{1}{2} \times 2200 \times 10^{-6} \times 9 \cdot 0^2$$

$$E = 0{\cdot}09 \ J$$

(ii) $I = \dfrac{V}{R} = \dfrac{9{\cdot}0}{100 \times 10^3} = 9{\cdot}0 \times 10^{-5} \ A$

p-n junctions

Poor descriptions of how an LED emits light

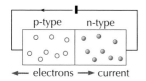

◄— electrons —► current

When a p-n junction is forward-biased, electrons move towards the conduction band of the p-type material and away from the n-type material. When the electrons drop from the conduction band to the valence band, energy is released in the form of a photon. This is the basis for light-emitting diodes (or LEDs).

Exam example 53

Q (b) Some cars use LEDs in place of filament lamps.

An LED is made from semiconductor material that has been doped with impurities to create a p-n junction.

The diagram represents the band structure of an LED.

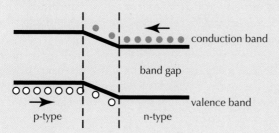

(i) A voltage is applied across an LED so that it is forward-biased and emits light.

Using **band theory,** explain how the LED emits light.   **3**

A Electrons move from n-type material towards the conduction band of the p-type material.

AND then

Electrons drop from the conduction band to the valence band.

AND as a result

A photon is emitted.

Note: 'Band theory' is in bold, indicating that there must be some reference in your answer to the conduction band and the valence band. If there is no reference to these then you cannot access any marks in this question.

Poor understanding of what it means to 'dope' a semiconductor material and the effect this has on conductivity

In a *pure* semiconductor material, such as silicon, all of the atoms are the same kind. A pure semiconductor has quite a high resistance, especially at low temperatures, and there is a low current when it is placed in a circuit.

When a small proportion of impurity atoms (such as indium or arsenic) are diffused into the semiconductor, its resistance decreases and its conductivity increases. Adding these impurity atoms is called 'doping' the semiconductor.

Silicon has four outer electrons. Doping using atoms that have five electrons in their outer shell increases the number of free *electrons* available for conduction. Four of these electrons fill up the valence band, leaving one electron in the conduction band. This is an *n-type* semiconductor.

Doping using atoms that have three electrons in their outer shell increases the number of positive *holes* available for conduction. The additional holes are just above the valence band, allowing electrons to move into them, and leaving spaces in the valence band to allow conduction. This is a *p-type* semiconductor.

Q A student writes the following statements about p-type semiconductor material.

    I Most charge carriers are positive.

    II The p-type material has a positive charge.

    III Impurity atoms in the material have three outer electrons.

Which of these statements is/are true?

A I only
B II only
C I and II only
D I and III only
E I, II and III

A p-type semiconductors are produced by doping using impurity atoms that have three electrons in their outer shells. Statement III is correct.

This produces gaps in the bonding structure called 'holes', which behave like positive charge carriers. There are now more holes than free electrons. Statement I is correct.

Statement II is incorrect because the added impurity atoms have the same number of positive protons as negative electrons and so are electrically neutral. This is because for every electron extra/less in a doping material there is an extra proton/or 1 less proton in the nucleus to maintain neutrality.

The answer is D.

Assignment

Researching Physics unit

The work you will carry out in preparation for writing your assignment will result in a half-unit award for Researching Physics.

The Researching Physics part of the course requires you to carry out:

- web-based research (1.1)
- a practical investigation (2.1 and 2.2).

For both the research and practical work, you must maintain a log book (record of work) with dates. The log book will be internally marked by your teacher as evidence for the unit award.

The log book should include a record of your web-based research (including references), all of your experimental results, ideas, problems you met and all the other day-to-day observations and data that you want to keep a record of.

Each entry in your log book should be dated so that your record of work is continually updated as you progress through the project.

Web-based research

The work in this part is to gather information that is relevant to the topic. This must include information on the underlying physics of your chosen topic.

When carrying out your research you must:

- make dated entries in your log book
- summarise relevant information in your own words
- note down the references (must be in enough detail for someone else to find them again)
- use at least two sources (two different websites).

If you download anything directly from the internet, it may suggest to the assessor that you have not understood the physics involved. This may be considered plagiarism unless you acknowledge the sources carefully. It is always best to put things in your own words to make sure, and demonstrate to the assessor, that you really understand them.

Practical investigation

When carrying out your practical investigation, your log book entries must have:

- a clear aim for the investigation
- a clear and detailed description of how the practical research investigation will be carried out, including safety considerations
- measurements to be made.

When carrying out the practical work:

- your teacher must observe you carrying out the experiments safely
- data must be recorded correctly using appropriate tables with headings and units.

Assignment overview

You will now use your log book, along with the data you have gathered during your research, to produce your assignment.

The first source of data should be from your practical investigation. The second source of supporting data should come from your research.

You need to select, process, present and analyse the information/data from the sources you have gathered and produce a written report under exam conditions. You cannot prepare a draft of your report.

As a guide, your report should be 800–1500 words long, excluding tables, charts and diagrams.

This table shows the number of marks available for each section of the report:

Criteria	Mark allocation
Aim(s)	1
Applying knowledge and understanding of physics	4
Selecting information	2
Processing and presenting data/information	4
Uncertainties	1
Analysing data/information	2
Conclusion(s)	1
Evaluation	3
Presentation	2

Assignment structure

Your report should be structured using the following headings.

Title: You must start your assignment with a title, such as 'The Physics of LEDs'.

Aim: The aim must clearly describe what is to be investigated. It should be a clear statement of what is to be investigated or measured, for example, 'to compare the energy of an earthquake with the maximum amplitude of the seismic waves'.

The aim should be stated separately from the title.

Underlying physics: Using the information collected from your log book during your web-based research, you should explain the underlying physics of your chosen topic.

Your response should include correct explanations of the topic researched, as well as the use of physics concepts and principles, such as refraction, internal resistance etc.

You should include formulae and calculations with quantities and units clearly defined.

You must demonstrate that you understand the physics involved. Copying directly from a reference website is not acceptable. *Explanations must be in your own words*. Information that is quoted from references in this section and then explained or expanded upon is acceptable.

Data and information: Provide raw data from two sources.

Source 1 – Data from your experiment

Here you must write a description of your experimental procedure, including:

- title
- aim
- method
- data (repeated measurements, appropriate range of values)
- uncertainties.

Source 2 – Selected information (your web-based supporting data)

This can take the form of:

- printouts of appropriate sections of webpages
- extracted tables, graphs, diagrams and text; statistical, graphical, numerical or experimental data
- published extracts of articles.

For the data/information to be sufficient, there should be enough of it to draw a conclusion from.

Processing and presenting data/information: In this section, you must present the data/information that you have processed from both your sources.

The processed data/information must be presented in appropriate format(s), for example:

- plotting graphs from tables
- populating a table from other sources
- summarising referenced texts.

At least one source must be processed into a graph, table or chart. There is no requirement to present data/information in different formats.

It must be clear where the data/information that you processed came from. It is helpful here to make reference to your raw data. For example, label 'graph 1, source 1'.

Check that you have included, as appropriate:

- suitable scales
- units
- headings
- labels
- line/curve of best fit.

When using graphing packages, both major and minor gridlines should be included. Points should be visible but not overly large. Lines or curves of best fit should be used.

Uncertainties: All reading uncertainties, including correct units, should be included for all measurements made.

If the reading uncertainty is the same for all measurements then a statement to this effect is all that is needed, for example, 'the reading uncertainty in each voltage is ±0.01 V'.

You should include calibration uncertainties for any meters used.

If repeated readings are taken then a random uncertainty in each set of readings should be calculated and included. A sample calculation should be included to show how the random uncertainty was calculated.

You should calculate the mean value for any repeated measurements.

Your final uncertainty will be the biggest percentage uncertainty.

Analyse data/information: Analysis will include interpreting data/information in the report to identify relationships. This may include further calculations.

For this section, ensure that you have:

- analysed your data
- compared your sources.

You may use either raw data/information (such as graphs or tables from the internet or journals) that you have included, or your processed data/information, or a combination of both.

Conclusion: You must clearly state the conclusion of your investigation. Your conclusion must relate to your aim and be supported by what you have found out in the course of your report.

Evaluation: This should include an evaluation of your individual sources and an evaluation of the investigation as a whole. You may include, for example:

- Validity of sources/robustness – are findings supported by other reputable sources? For experimental work, describe how the key variables were controlled.
- Reliability of data/information – from a scientific journal, sample size, repeated results (to reduce the random uncertainty for example).
- Evaluation of experimental procedures:
  - adequacy of repetition – repeated results taken
  - limitations of equipment used
  - improvements to experiment (such as greater range of currents, repeat experiment with different colours of LED)
  - sources of uncertainty (reading, calibration, random)

References: At the end of your report, you must state your references. Although references may appear within the body of the report, they must also appear at the end of the report.

- Give the title and the aim of your experiment/practical activity.
- Record the sources (full URLs) you have used, with enough detail to allow someone else to find them:
  - References of websites must be complete URL addresses – www.bbc.co.uk is not acceptable.
  - References of textbooks must include title, author, page number and either ISBN number or version/edition.

- References of journals must include journal title, author, volume, year, and page number.
- At least two references must be given.

Areas of difficulty

Listed below are some of the common areas of difficulty candidates experience when completing the assignment.

Level of demand

Care should be taken when selecting your assignment topic. The topic should relate to a key area from within the Higher Physics course, not to courses at other levels. A topic that relates to the National 5 or Advanced Higher course will not score highly.

Uncertainties

Many candidates neglect to include a sample calculation, with formula, for random uncertainties.

All uncertainties must include units where appropriate.

All reading uncertainties must be included.

Analysis

This section should involve an extraction of information from your findings. This can be done through:

- the calculation of a constant
- a description of how uncertainties have influenced the data
- discussion of systematic uncertainties in data
- calculation of a gradient (including an explanation of the significance of the value).

Evaluation

This section should contain three separate evaluative comments. Each statement should be supported by appropriate justification.

Aim/Conclusion

When selecting the aim of your assignment, you should try not to choose one that is overly complex. The conclusion has to address all aspects of the aim, so it becomes difficult to access the marks for the conclusion if the aim is not concise.

Title

Your title should reflect what is being investigated. 'Higher Assignment' is not an appropriate title.

Processing and presenting information

It is useful to the marker if you clearly label raw data and processed data, such as 'Source 1, raw data'; 'Source 2, processed data'.

References

References should be **the final item** in the report. You will lose several marks if you do not list your references at the end.

Notes

Notes

Notes

Notes

Notes